STELLAR INSTABILITY AND EVOLUTION

INTERNATIONAL ASTRONOMICAL UNION
UNION ASTRONOMIQUE INTERNATIONALE

SYMPOSIUM No. 59

HELD AT MOUNT STROMLO, CANBERRA, AUSTRALIA,
16–18 AUGUST, 1973

STELLAR INSTABILITY AND EVOLUTION

EDITED BY

P. LEDOUX AND A. NOELS

Institut d'Astrophysique, Université de Liège, Cointe-Ougrée, Belgium

AND

A. W. RODGERS

Mount Stromlo and Siding Spring Observatory, Australia

D. REIDEL PUBLISHING COMPANY

DORDRECHT-HOLLAND / BOSTON-U.S.A.

1974

Published on behalf of
the International Astronomical Union
by
D. Reidel Publishing Company, P.O. Box 17, Dordrecht, Holland

All Rights Reserved
Copyright © 1974 by the International Astronomical Union

Sold and distributed in the U.S.A., Canada, and Mexico
by D. Reidel Publishing Company, Inc.
306 Dartmouth Street, Boston,
Mass. 02116, U.S.A.

Library of Congress Catalog Card Number 74-80520

Cloth edition: ISBN 90 277 0479 1
Paperback edition: ISBN 90 277 0480 5

No part of this book may be reproduced in any form, by print, photoprint, microfilm,
or any other means, without written permission from the publisher

Printed in The Netherlands by D. Reidel, Dordrecht

TABLE OF CONTENTS

PREFACE — VII

LIST OF PARTICIPANTS — IX

PART I / SURVEY OF THE PROBLEMS IN RADIAL STELLAR INSTABILITY AND ITS RELATION TO STELLAR EVOLUTION

I. IBEN JR. / The Theoretical Situation — 3
O. J. EGGEN / The Observational Evidence — 35

PART II / PULSATION AND THE YOUNG DISC POPULATION

A. N. COX / Recent Progress in Linear and Non-Linear Calculations of Radial Stellar Pulsation — 39
R. S. STOBIE / Population I Cepheids – Theory and Observation — 49
S. B. PARSONS / The Temperature Scale for Classical Cepheids — 55
R. A. BELL / $uvby$ Colours of Cepheid Variables — 57
K. BAHNER and L. N. MAVRIDIS / Period Variations in Galactic Cepheids (*Title only*)
A. N. COX, J. E. TABOR, and D. S. KING / Non-Pulsating Stars and the Population I and II Instability Strips — 59
M. BREGER / Pulsation Modes and Period-Luminosity-Color Relations for the Short-Period δ Scuti Stars — 61
M. S. BESSELL / The Abundances and Gravities of the δ Scuti, AI Velorum and RR Lyrae Stars — 63
D. H. MCNAMARA and W. R. LANGFORD / The Dwarf Cepheids — 65
R. R. SHOBBROOK and R. S. STOBIE / Multiple Periodicities in δ Scuti Stars — 67
R. R. SHOBBROOK / β Canis Majoris Stars in the (β, $[c_1]$) Plane — 69
A. N. COX and J. E. TABOR / A Search for β Cephei Pulsation in Extreme Composition Models — 73

PART III / LARGE MASS STARS, STABILITY IN SUPERGIANTS, CRITICAL MASSES

J. C. B. PAPALOIZOU and R. J. TAYLER / Non-Linear Instability of Stars with $M > 100 \ M_\odot$ — 77
A. MAEDER and F. RUFENER / On the Variability of Supergiants of types B–G — 81
P. S. OSMER / Recent Studies of the Brightest Stars in the Magellanic Clouds — 85
W. BUSCOMBE / Variations in the Line Spectrum of A-Type Supergiants — 87

PART IV / HALO AND OLD DISC POPULATIONS

A. *Red Variables and Evolution on the Giant Branches*

M. W. FEAST / Red Variables of the Old Disc and Halo Populations	93
P. R. WOOD / Dynamical Models of Asymptotic-Giant-Branch Stars	101

B. *Short Period Variables, Horizontal and Asymptotic Branch Evolution and Planetary Nuclei*

P. DEMARQUE / Variable Stars and Evolution in Globular Clusters	105
J. A. GRAHAM / The Characteristics of the Field RR Lyrae Stars in the Magellanic Clouds	107
A. MAEDER / On Some Peculiar Features in the Sequences of the Old Open Star Clusters	109
J. KATZ, R. MALONE, and E. E. SALPETER / Models for Nuclei of Planetary Nebulae and Ultraviolet Stars	113

PART V / ERUPTIVE AND EXPLOSIVE VARIABLES

E. SCHATZMAN / Explosive and Eruptive Stars	117
B. L. WEBSTER / On the Binary Nature of the Slow Nova, RR Telescopii	123
B. WARNER / Non-Radial Pulsations of Dwarf Novae	125
J. B. HUTCHINGS / Tests of the Binary Hypothesis of Novae through Nova Nebulae Observations	127
J. G. DUTHIE, B. RENAUD, and M. P. SAVEDOFF / Proposed Search for Fast Variables	129

PART VI / INSTABILITY MECHANISMS

A. *Non-Radial Pulsations – Magnetic Fields*

P. LEDOUX / Non-Radial Oscillations	135
K. H. SCHATTEN / The Sun as a Pulsating Rotating Star	175
R. J. TAYLER / Instabilities of Magnetic Fields in Stars	177
J. C. KEMP / Magnetic Fields in X-Ray Binary Systems	179
R. K. KOCHHAR and S. K. TREHAN / The Oscillations and the Stability of Rotating Masses with Magnetic Fields	181

B. *Rotation – Close Binaries*

J.-P. ZAHN / Rotational Instabilities and Stellar Evolution	185
P. SMEYERS / Perturbation of the Non-Radial Oscillations of a Gaseous Star by an Axial Rotation, a Tidal Action or a Magnetic Field	197
J. B. HUTCHINGS / Mass-Ratio Determination in Contact Binaries	199
INDEX OF NAMES	201

PREFACE

The XXth meeting of the IAU in Australia in 1973 made the venue for the IAU Symposium No. 59 on Stellar Instability and Evolution, at Mount Stromlo Observatory on August 16–18, a very appropriate one. Many of the current and former staff of Mount Stromlo Observatory (operated by the Australian National University) have specialized in the study of variable stars and it was with considerable pleasure that Mount Stromlo Observatory accepted the responsibility of hosting and making the local arrangements for IAU Symposium No. 59.

The Scientific Organizing Committee was particularly active in formulating the program and comprised Drs N. Baker, P. Demarque, M. Feast, G. Herbig, I. Iben, P. Ledoux, J. Ostriker and E. Schatzman. The aim of the Committee was to integrate the review and contributed papers on the particular instability mechanisms involved, their observational manifestations and their relation to the internal structure of the star as inferred from its evolutionary history.

The Local Organizing Committee consisted of Miss P. Petrie and A. W. Rodgers. Miss Petrie made all of the local arrangements and she was ably helped during the running of the Symposium by Mesdames P. Lea, V. Bloxham and J. Hinchey and Messrs. C. Smith, H. Butcher, J. Mould, R. Gingold and G. da Costa. The Symposium, officially opened by the Acting Vice Chancellor of the Australian National University, Professor N. Dunbar, was held in the lecture theatre of the Observatory some ten miles from Canberra, and theorists were confronted with the cold realities of observing (and Australian meat pies) by having lunch served to them in the spacious dome of the 26-inch refractor on Mount Stromlo.

According to the IAU publishing rules, the invited papers are printed in full extent, while only abstracts of the contributed papers are given.

As far as the discussions are concerned, the best had to be made of the transcription of a tape recording and of whatever written comments reached the editors. The latter present their apologies for the imperfections and omissions, but they hope that, in the present form, the printed discussions reflect the liveliness and the interest of the debates.

LIST OF PARTICIPANTS

Aizenman, M. L., JILA, University of Colorado, Boulder, Colo. 80302, U.S.A.
Bateson, F. M., 18, Pooles Road, Greerton Tauranga, New Zealand
Bell, R. A., Dept. of Physics and Astronomy, University of Maryland, College Park, Md. 20742, U.S.A.
Bessell, M. S., Mount Stromlo and Siding Spring Observatory, Australian National University, Private Bag, Woden P.O., A.C.T. 2606, Australia
Breger, M., Dept. of Astronomy, The University of Texas at Austin, Austin, Tex. 78712, U.S.A.
Buscombe, W., Dept. of Astronomy, Northwestern University, Evanston, Ill. 60201, U.S.A.
Clement, M. J., Dept. of Astronomy, University of Toronto, Toronto, Ontario, Canada M 5S 1A7
Cogan, B. C., Physics Dept., Virginia Polytechnic Inst., Blacksburg, Va. 24061, U.S.A.
Cox, A. N., Los Alamos Scientific Laboratory, P.O. Box 1663, N.M. 87544, U.S.A.
Demarque, P., Yale University Observatory, Box 2023, Yale Station, New Haven, Conn. 06520, U.S.A.
Edwards, P. J., University of Otago, New Zealand
Eggen, O. J., Mount Stromlo and Siding Spring Observatory, Australian National University, Private Bag, Woden P.O., A.C.T. 2606, Australia
Faulkner, D., Mount Stromlo and Siding Spring Observatory, Australian National University, Private Bag, Woden P.O., A.C.T. 2606, Australia
Feast, M. W., Science Research Council, Radcliffe Observatory, P.O. Box 373, Pretoria, South Africa
Graham, J. A., Cerro Tololo Inter-American Observatory, Casilla 63-D, La Serena, Chile
Hartwick, F. D. A., Dept. of Astronomy, University of Victoria, Victoria, B.C., Canada
Hutchings, J. B., National Research Council of Canada, Dominion Astrophysical Observatory, RR7, Victoria, B.C., Canada
Hyland, H. R., Mount Stromlo and Siding Spring Observatory, Australian National University, Private Bag, Woden P.O., A.C.T. 2606, Australia
Iben, I. Jr., Dept. of Astronomy, University of Illinois at Urbana-Champaign, Urbana, Ill. 61801, U.S.A.
Irwin, J. B., Dept. of Earth and Space Science, Newark State College, Union, N.J. 07083, U.S.A.
Jorgensen, H. E., Copenhagen University Observatory, Øster Voldgade 3, DK-1350, Copenhagen K., Denmark

Kemp, J. C., Dept. of Physics, University of Oregon, Eugene, Ore. 97403, U.S.A.
Kirshner, R., Hale Observatories, California Institute of Technology, Pasadena, Calif. 91109, U.S.A.
Kuhi, L. V., Institut d'Astrophysique, 98 bis, boulevard Arago, Paris 14e, France
Larsson-Leander, G., Lund Observatory, Svanegatan 9, S-22224, Lund, Sweden
Ledoux, P., Institut d'Astrophysique, University of Liège, Cointe-Ougrée, B-4200, Belgium
Lesh, J. R., Dept. of Physics and Astronomy, University of Denver, Denver, Colo. 80210, U.S.A.
Lockwood, G. W., Lowell Observatory, Flagstaff, Ariz. 86001, U.S.A.
McNamara, D. H., Physics Dept., Brigham Young University, Provo, Utah 84602, U.S.A.
Maeder, A., Observatoire de Genève, 1290 Sauverny, Switzerland
Mathewson, D., Mount Stromlo and Siding Spring Observatory, Australian National University, Private Bag, Woden P.O., A.C.T. 2606, Australia
Mavridis, L. N., Dept. of Astronomy, University of Thessaloniki, Thessaloniki, Greece
Mestel, L., Dept. of Mathematics, The University, Manchester 13, U.K.
Osmer, P. S., Cerro Tololo Inter-American Observatory, Casilla 63-D, La Serena, Chile
Parsons, S. B., Dept. of Astronomy, The University of Texas at Austin, Austin, Tex. 78712, U.S.A.
Przybylski, A., Mount Stromlo and Siding Spring Observatory, Australian National University, Private Bag, Woden P.O., A.C.T. 2606, Australia
Rodgers, A. W., Mount Stromlo and Siding Spring Observatory, Australian National University, Private Bag, Woden P.O., A.C.T. 2606, Australia
Salpeter, E. E., Laboratory of Nuclear Studies, Cornell University, Ithaca, N.Y. 14850, U.S.A.
Sargent, W. L., Astronomy Dept., California Institute of Technology, Pasadena, Calif. 91109, U.S.A.
Savedoff, M. P., Mees Observatory, University of Rochester, Rochester, N.Y. 14627, U.S.A.
Schatten, K. H., Victoria University of Wellington, New Zealand
Schatzman, E. DAF, Observatoire de Paris-Meudon, 92190-Meudon, France
Shobbrook, R. R., Dept. of Astronomy, The University of Sydney, Sydney 2006, N.S.W., Australia
Sinvhal, S. D., Uttar Pradesh State Observatory, Naini Tal, India
Smeyers, P., Astronomisch Instituut, Katholieke Universiteit Leuven, Naamsestraat 61, B-3000 Leuven, Belgium
Stobie, R. S., Mount Stromlo and Siding Spring Observatory, Australian National University, Private Bag, Woden P.O., A.C.T. 2606, Australia
Tayler, R. J., Astronomy Center, University of Sussex, School of Math. and Phys. Sciences, Falmer, Brighton BNI 9 QH, U.K.

Trehan, S. K., Centre for Advanced Study in Mathematics, Panjab University, Chandigarh – 160014, India

Trodahl, H. J., Wellington, New Zealand

Underhill, A. B., Goddard Space Flight Center, Greenbelt, Md. 20771, U.S.A.

van Agt, S., Astronomical Dept. of the University of Nijmegen, Nijmegen, The Netherlands

van Albada, T. S., Kapteyn Astronomical Laboratory, Hoogbouw WSN P 800, Groningen 8002, The Netherlands

Van den Borght, R., Mathematics Department, Monash University, Melbourne, Australia

Van Woerden, H., Kapteyn Astronomical Laboratory, Hoogbouw WSN P 800, Groningen 8002, The Netherlands

Vardya, M. S., Astrophysics Division, Tata Institute of Fundamental Research, Colaba, Bombay 400005, India

Warner, B., Dept. of Astronomy, University of Cape Town, Rondebosch, C.P., South Africa

Webster, B. L., Science Research Council, Radcliffe Observatory, P.O. Box 373, Pretoria, South Africa

Wielen, R., Astronomisches Recheninstitut, Mönchhofstr. 12–14, 6900 Heidelberg, F.R.G.

Wolstencroft, R. D., Institute of Astronomy, University of Hawaii, 2840 Kolowalu Street, Honolulu, H.I. 96822, U.S.A.

Wood, F. B., Dept. of Physics and Astronomy, University of Florida, Gainesville, Fla. 32601, U.S.A.

Wood, P. R., Mount Stromlo and Siding Spring Observatory, Australian National University, Private Bag, Woden P.O., A.C.T. 2606, Australia

Wright, K. O., Dominion Astrophysical Observatory, RR 7, Victoria, B.C., Canada

Zahn, J.-P., Observatoire de Nice, Le Mont-Gros, 06 Nice – France

PART I

SURVEY OF THE PROBLEMS IN RADIAL STELLAR
INSTABILITY AND ITS RELATION
TO STELLAR EVOLUTION

THE THEORETICAL SITUATION

ICKO IBEN, Jr.

University of Illinois, Champaign – Urbana, Ill., U.S.A.

1. Preliminary Remarks

The domain of this symposium is so wide – essentially anything at all having to do with stars – that it is impossible to do more than cursory justice to even a small portion of the matters to be discussed. My contribution will therefore be limited primarily to a discussion of the status of theoretical work bearing on the behavior of stars that evolve through the classical instability strip that extends from the region of Cepheids to the domain of RR Lyrae stars. Discussion of other extremely important variable stars such as cataclysmic variables, Mira and irregular variables, flare stars, β-Canis Majoris stars, and δ-Scuti and small amplitude variables will here be mentioned only in passing; presumably, most of these stars will be discussed at length by other speakers at this symposium. Further little attention will be paid to the thermal instability that is initiated in the helium-burning region of a double-shell-source star and to current thinking about the progenitors of type I and type II supernovae; presumably, these topics will be discussed in Warsaw in the symposium on advanced stages of evolution.

2. An Overview

I will first attempt to summarize: (1) what has been known for a long time, certainly by the time of the last IAU Symposium on Stellar Evolution and Pulsation, (2) what has happened since this last symposium, and (3) what some of the striking deficiencies in the theory are, that can and ought to be remedied.

2.1. What's old?

(1) Most of the regular variables of large amplitude are in the core helium-burning stage. Classical Cepheids are stars of intermediate mass (say 3–16 $M_\odot$); RR Lyrae stars are of low mass (say ~ 0.5–$0.7\ M_\odot$). Both pulsation theory and evolution theory, when compared with the appropriate observations, give these results. The beauty of the agreement is that the comparisons are, to a large extent, independent of one another. Pulsation theory deals primarily with the outer envelope of the star (temperatures below 10^6 K) whereas evolution theory deals in large part with the stellar core where the essential input physics is quite different from that in the envelope.

(2) The principal driving region for most regular variables is the He II ionization zone, although the hydrogen ionization zone can play a critical role, particularly in determining the mode in which a star will pulsate.

(3) The observational instability strip extends to luminosities considerably below

those of RR Lyrae variables, thus extracting from an embarrassing situation theoreticians who are not able to easily stabilize such stars. Among the stars near the main sequence are the δ-Scuti variables, many of which may be burning hydrogen either at the center or at the edge of an inert helium core.

(4) The relationship between P, T_e, L, and M that is obtained by either linear or non-linear analysis is perhaps one of the most secure relationships in astrophysics.

(5) Blue edges for pulsation in all modes are sensitive to the envelope helium abundance as well as to mass. Lifetimes in different evolutionary phases and locations in the HR diagram are both sensitive to the envelope helium abundance. On comparison with the observations, these sensitivities both suggest a high helium abundance for population II variables.

(6) The fact that, at low T_e, pulsation in the fundamental mode seems to be favored over pulsation in the first harmonic mode and that, at high T_e, the reverse is true, leads to the suspicion that there may be a roughly composition-independent, mass-independent relationship between a 'transition' period and luminosity. A guess at this relationship permits one to estimate distances to RR Lyrae stars near the galactic center and in globular clusters. Stellar evolution calculations lead to an independent relationship between luminosity and composition that may also be used to estimate distances to RR Lyrae stars. Both methods (evolution and pulsation) agree on the sense and magnitude of the difference in luminosity between variables in clusters of intermediate metal content and variables in clusters of very low metal content.

(7) The red edge of the instability strip found in nature is probably connected with convection, which quenches the driving mechanism. Theory, however, cannot yet predict the location of a red edge.

(8) β-Cephei stars are probably near the overall contraction phase following the exhaustion of hydrogen in the core. There is the possibility that non-radial pulsation modes are involved and that a resonance coupling between such modes and rotation and/or with radial modes may occur. However, an excitation mechanism has not yet been demonstrated.

(9) Novae are probably binary systems, one member of which is a white dwarf and the other member of which is a red giant that transfers mass to its companion. When the temperature at the base of the newly deposited hydrogen layer exceeds a critical value, an explosion takes place which lifts off the new layer.

(10) Supernovae models of a non-exotic, non *ad-hoc* nature are almost nonexistent. Only one model has respectable antecedents – the carbon detonation model. Unfortunately, this model suffers from three serious flaws: no remnant; too much iron produced; possibly not enough energy liberated.

(11) Pulsars have been identified and, for the Crab, the only viable theoretical explanation is that of a rotating neutron star.

(12) Long period irregular variables are ubiquitous in the redder regions of the HR diagram. Scandalous to relate, neither the excitation mechanism nor the evolutionary status of the interior is understood.

(13) In young clusters, flare stars are common and unpredictable. Flares are

thought to be a surface phenomenon associated perhaps with the annihilation of magnetic field.

(14) A thermal instability occurs in the helium shell of stars that have exhausted helium at their centers. The amplitude of this instability increases with time. A convective shell that appears within the helium-burning shell and extends outward toward the hydrogen-helium discontinuity grows to a maximum during the peak of a thermal swing and may lead to mixing of hydrogen into the helium-burning region and/or mixing of helium-burning products into the hydrogen-rich region. In either case, s-process elements may be formed and eventually brought to the surface. The first property of the thermal instability – the increase in amplitude with each pulse – may be a clue to the formation of some planetary nebulae. The second property – the formation of an 'intershell' convective region – may be responsible for the peculiar abundances in carbon stars and the like.

(15) For masses in the neighborhood of 60 $M_\odot$, main sequence stars become unstable to radial pulsation driven by nuclear reactions in the core. Attempts at following the non-linear behavior of such objects suggest that, within the main sequence lifetime, pulsation amplitude does not grow without limit and that, therefore, pulsational instability is not the reason for the paucity of high mass stars. The reason for this paucity must reside in the conditions prevailing during star formation.

(16) U-Gem, W-Ursa Majoris stars, dwarf novae, and classical novae all fall into a regular scheme. In this scheme the separation between members of a binary pair figures as a crucial parameter.

2.2. What's new?

This, of course, is the subject of our conference, and we eagerly await tidings of new advances. Some advances are already in the literature and may therefore be reviewed.

(1) Extensive use has been made of P, L, T_e, M relationships and blue edge relationships to estimate the bulk properties of Cepheids and RR Lyrae stars. The uncertainties in these relationships have been extensively examined and appreciated.

(2) Evolutionary studies have clarified that many population II variables with period greater than a day are stars with a carbon-oxygen core that are burning nuclear fuel in two shells. BL-Herc stars are evolving through the instability strip on a nuclear burning time scale (in the suprahorizontal branch or asymptotic giant branch stage); W-Virginis stars are evolving through the instability strip on a thermal time scale (looping back from the asymptotic giant branch).

(3) The tentative relationship between a transition period and luminosity (discussed in 2.1.6) has fallen under a cloud. It appears that the relationship is really only a rough estimate of the location of a transition region where pulsation begun in either of two modes – fundamental or first harmonic – will persist in the initial mode for many periods. It possibly cannot be trusted to give luminosity differences as small as those which occur between RR-Lyrae stars in clusters of different composition.

(4) A powerful new technique for calculating full amplitude motion directly has been invented. Instead of integrating forward in time by brute force, a periodic solution

is sought by relaxing in time as well as in space. The final full amplitude motion can then be tested for stability. In principle one can thereby determine which of two unstable modes may survive over a long period of time. The new technique may remove the cloud now hanging over the relationship between a transition period and luminosity or place this relationship more permanently in limbo. Preliminary results suggest that the cloud will be reduced but not completely eliminated.

(5) The location of β-Cephei stars has been more carefully delineated. This location appears to be a band parallel to the main sequence and along the red edge of the main sequence band. The evidence is therefore rather compelling that β-Cephei stars are either in or near the overall contraction phase following the exhaustion of hydrogen at the center. Still, a completely acceptable excitation mechanism has not yet been identified.

(6) Dynamic studies of nova-like outbursts have been conducted on the assumption that the appropriate initial model consists of a white dwarf accreting hydrogen-rich matter from a red-giant companion. It develops that much of the power for the late stages of the outburst comes from the β-decay of ^{13}N and ^{15}O and that theoretical bolometric light curves are similar to those observed for nova outbursts in the visual. However, most of the energy associated with nova outbursts may not be in the visual.

(7) The carbon-detonation model of supernovae has been transformed into the carbon-fizzle model. Instead of exploding, the core now is thought to lose energy via Urca-process neutrinos at the edge of a convective core whose extent is determined by the threshold energy for β^- captures on ^{23}Na ($\to$ ^{23}Ne). Conveniently, ^{23}Na is one of the products of reactions that occur during carbon burning. As a consequence of the Urca energy drain, the incipient detonation may be quenched. As carbon burning proceeds quietly, neutron-rich elements become ever-more abundant (the electron molecular weight μ_e increases) until the Chandrasekhar limit ($\propto M_\odot/\mu_e^2$) exceeds the mass in the electron-degenerate carbon-oxygen core. The core may then collapse into a pulsar. It is possible that the beam from the pulsar may fill up the cavity between pulsar and envelope with electromagnetic energy which then expels the envelope. Thus, we are left with a condensed remnant that can act as a reservoir for large quantities of energy and with an expelled envelope that contains *no* new elements to speak of. For the heavy elements, it appears that we may be forced to rely on massive stars that don't develop instabilities until an iron or nickel core is formed.

(8) It is becoming increasingly probable that type I supernovae may initially be binary systems with rather long periods ~ 1–10 yr. It is conjectured that one component has an initial main sequence mass on the order of 2–$3\frac{1}{2}$ $M_\odot$, the other an initial main sequence mass less than or equal to about 0.8 $M_\odot$. The time scale for the system is determined by the lighter star. The more massive star sheds mass due to radiation-pressure induced mass flow and becomes a carbon-oxygen white dwarf. Most of the mass is lost from the system. After $\sim 10^{10}$ yr, the secondary transfers mass to the white dwarf which heats up, reaches the Chandrasekhar limit and (maybe) explodes. Nice features of the conjectured model are a high degree of symmetry, the near uniqueness

of the immediate supernova progenitor, and an explosion energy that is not masked by a massive envelope (as may be the case for type II supernovae).

(9) Pulsars in binary systems of very short period provide a new puzzle. If the X-ray pulsars are indeed neutron stars, how could the companion possibly survive the explosion that presumably precedes the formation of a neutron star remnant? Maybe explosions don't always accompany the formation of neutron stars.

(10) Long period irregular variables may owe their variability to random convective motions that have their origin in the thermally unstable region defined by the helium and hydrogen-burning shells.

(11) Further exploration of s-process element formation under conditions anticipated in thermally unstable regions speaks more and more strongly for interplay between the intershell region and the surface. The key factors in achieving the appropriate relative abundances seem to be (a) mixing between hydrogen and products of helium burning and (b) repetition of the mixing process for hundreds of cycles. Further calculation of thermal instabilities has produced possibly consistent, but also possibly contradictory results – low mass stars may not mix; higher mass stars can't seem to avoid it. The observations (FG Sagittae) suggest that mixing may take place in the deep interior of low mass stars, but appearance of the exotic products at the surface may not occur until after the ejection of a planetary nebula.

(12) The rapid oscillations that occur following the nova outburst appear to be explicable only in terms of non-radial modes in a white dwarf.

(13) Some progress in understanding the structure of contact binaries has been made, but estimates of evolutionary behavior by different authors give contradictory results.

2.3. WHAT SHOULD BE DONE?

(1) In my mind, one of the most vital elements in the comparison between observation and theory has received insufficient attention – the conversion between observational quantities and theoretical properties. For example, how does the conversion between $B-V$ and $\log T_e$ depend on composition? Until this conversion has been fully explored, we cannot properly understand the differences between Cepheids in our Galaxy, the SMC, the LMC, and Andromeda. Nor can we fully appreciate the dispersion in properties of Cepheids in our own Galaxy. Again, what are $\langle B-V \rangle$ and $\langle B \rangle - \langle V \rangle$ for a model pulsating star characterized by a specific $\log T_e$ at zero amplitude? Without a thoroughgoing study of these transformations we cannot make rigorous statements about masses. Exploration of these transformations is a task for theoreticians.

(2) We need a more complete study of the composition dependence of evolutionary tracks for stars of intermediate mass during core helium burning. Only then can we understand differences between Cepheids in the SMC, LMC, Andromeda, and our Galaxy.

(3) What is the excitation mechanism for β-Cephei stars? Is it sporadic convective motions that arise in massive stars during the phase of overall contraction when a

large burst of energy is emitted by the contracting core? It may be significant that this burst of energy does produce a thick convective shell outside of the hydrogen-exhausted core.

(4) We need a convincing demonstration of the existence or non-existence of a transition period. Is there a unique period at any given L and M? If so, does it depend on composition and mass? Is there only a transition region of finite width? That is, is there a hysteresis effect so that a star will continue to pulsate in the mode in which it began as it passes through the transition *region* where both modes are possible?

(5) We need an explanation of the observed drifts in period which occur on time scales short compared to the drifts brought about by evolutionary changes. Are the rapid drifts due to perturbations that couple rotation and pulsation? Do magnetic fields play a role?

(6) Do we really understand how to estimate the effects of convective overshoot and of semi-convection at the outer boundaries of fully convective cores? The fact that the helium abundance in the envelopes of horizontal branch stars that is given by pulsation theory plus observations is significantly larger than that given by evolution theory (with overshoot and semi-convection treated in a particular way) plus observation suggests that something is wrong.

(7) The evolution of contact binaries needs further attention. Currently, theoretical results are contradictory.

(8) The possibility of supernova explosions connected with a carbon-oxygen core of mass near 1.4 $M_\odot$, whether inside of a star of intermediate mass or adjacent to a light donor that sends it over the brink, needs further extensive exploration.

(9) What is the cause of the planetary nebula phenomenon? An extremely large amplitude thermal instability initiated in nuclear burning regions? Radiation-driven mass loss when stellar luminosity exceeds a high, critical value? An exceedingly large amplitude envelope pulsation driven by a helium or hydrogen ionization zone? Or do all three mechanisms play a role?

3. Classical Cepheids

3.1. Gross features

Since the firm identification of main sequence stars as objects in the phase of core hydrogen burning, perhaps the most significant advance in our understanding of stars has been the firm identification of classical Cepheids and cluster variables as objects in the core helium-burning phase. This identification has provided a most convincing demonstration of the qualitative correctness of the theoretical framework of stellar structure and pulsation.

In Figure 1, the shaded areas represent where, according to evolutionary calculations (see, e.g., Iben, 1967a, b), stars of population I composition spend most of their time as consumers of nuclear fuel. Over 80% of a theoretical model's total lifetime is spent in the pure hydrogen-burning phase. Another 10 to 20% is spent burning both helium and hydrogen. The location of models in more advanced phases is on the giant branch. All details of evolutionary tracks have been suppressed in order to focus on the major qualitative intersection of evolution and pulsation theory: After its residence on the main sequence, a population I star of intermediate mass spends most of its time in a band of finite width that is widely separated from the main sequence band and is roughly orthogonal to a narrow band defined by pulsation theory as a region within which the stellar envelope is unstable to pulsation in the fundamental radial mode.

An immediate consequence of the theoretical results is that Cepheids should be limited to a finite range in luminosity and period. That this is in fact the case is beautifully demonstrated in Figure 2 where the distribution in apparent magnitude is shown for Cepheids in M31 (data from Baade and Swope, 1965).

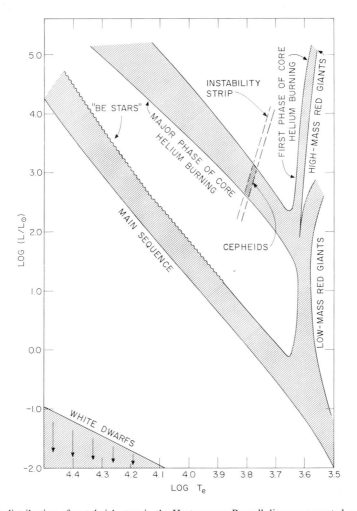

Fig. 1. The distribution of metal-rich stars in the Hertzsprung-Russell diagram expected on the basis of model calculations (from *Science* **155** (1967), 785).

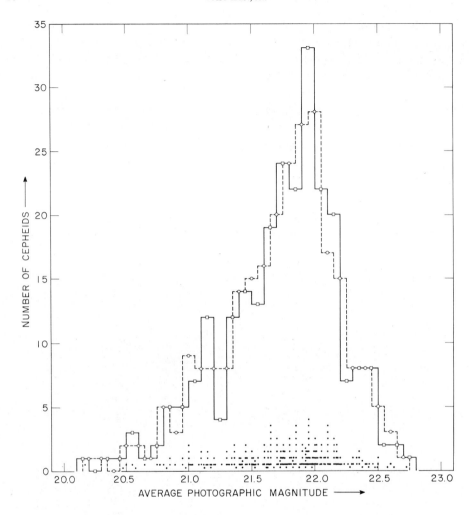

Fig. 2. Number vs magnitude relationship for Cepheids in Field III of M31 (after Baade and Swope, 1965). Unconnected points indicate the occurrence of a Cepheid at the designated magnitude (arithmetic mean of maximum and minimum magnitudes). For them, the vertical scale has no significance. The solid histogram is obtained by counting Cepheids in 0.1-mag intervals centered at 20.05, 20.15, etc. The dashed histogram is obtained by counting Cepheids in 0.1-mag. intervals centered at 20.1, 20.2, etc.

A second confirmation of the basic qualitative features of the theory is provided by the study of young clusters that contain stars massive enough to have reached the core helium-burning band (within the lifetime of the cluster). From the location of the cluster main sequence, one can estimate the absolute luminosity of the helium-burning stars and from the observed color one can estimate surface temperature. Theory suggests that, in younger clusters (that contain more massive stars), helium-burning stars will in general be brighter and bluer than such stars in older clusters. This trend is very nicely corroborated by clusters such as NGC2164 (Hodge and Flower, 1973), NGC 1866 (Arp 1967, Arp and Tackeray, 1967), and NGC1831 (Hodge 1963).

3.2. UNCERTAINTIES IN THE THEORY AND IN THE INTERPRETATION OF OBSERVATIONS

Although gross features of evolution and pulsation theory are in beautiful conformity with the gross features of the observations, attempts at detailed comparisons are plagued by large uncertainties. One may classify the uncertainties in three groups: (i) uncertainties in stellar structure; (ii) uncertainties in pulsation theory; and (iii) uncertainties attending an interpretation of observed properties.

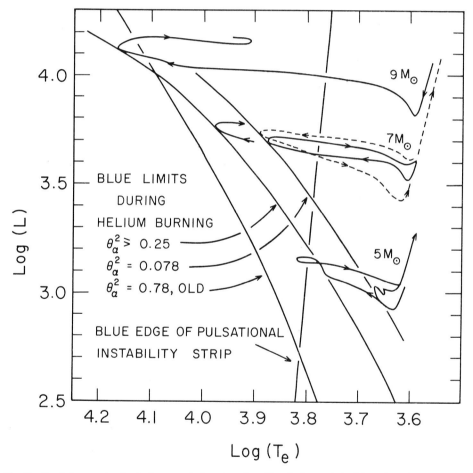

Fig. 3. Evolutionary tracks during core helium burning, blue limits, and the blue edge of a theoretical instability strip for radial pulsation in the fundamental mode. Initial composition parameters are $X=0.7$, $Y=0.28$, $Z=0.02$. Three choices of reduced width θ_α^2 are represented. For the $5M_\odot$ models, $\theta_\alpha^2=0.078$ and 0.25; for the $7M_\odot$ models, $\theta_\alpha^2=0.078$ and 0.78; for the $9M_\odot$ model, $\theta_\alpha^2=0.25$. The blue limits connect, for different masses, the bluest points reached during the major core helium-burning phase. A blue limit defined by an earlier set of models (see Iben, 1967a, b) is also shown.

(i) The uncertainties in stellar structure and evolution may be centered about the information in Figure 3, where tracks during the helium-burning phase for stars of 5, 7, and 9 $M_\odot$ are shown (Iben, 1972). Tracks have been constructed for specific choices of the input physics. An indicator of uncertainties is given by the 'blue limits', each limit being defined by connecting (for models of different mass) the highest surface temperature reached by each model during the core helium-burning phase.

The blue limits labeled '$\theta_\alpha^2 \gtrsim 0.25$' and '$\theta_\alpha^2 = 0.78$, old' are defined by models that differ *only* with regard to opacities. On examining the intersection of these two blue limits with the blue edge of the instability strip (for a similar composition), it is apparent that the two choices for opacity suggest mean Cepheid luminosities that differ by over 1.25 mag.! Thus, the choice of opacity excercises considerable influence over predicted Cepheid properties.

The only difference between models that define blue limits labeled '$\theta_\alpha^2 = 0.078$' and '$\theta_\alpha^2 \gtrsim 0.25$' is in the effective cross section for the reaction $^{12}C\,(\alpha, \gamma)\,^{16}O$ that occurs in the convective core of the models. Model stars constructed with a larger cross section evolve for a longer time and to a larger surface temperature than do model stars constructed with a smaller cross section. Laboratory nuclear physics experiments cannot yet exclude either choice of cross section. Thus, uncertainties in the nuclear physics introduce a large uncertainty in the predicted properties of Cepheids.

The treatment of mixing in the convective core of model stars introduces still further uncertainties. As helium turns into carbon and oxygen in this core, the opacity at the outer edge of the core grows larger than the opacity just outside the core (Castellani *et al.*, 1971a, b). The incipient discontinuity in opacity leads to finite convective velocities at the formal core edge and hence to overshooting. The consequence is that the effective region in which matter is mixed is larger than is the case when overshoot is neglected. The concomitant increase in nuclear fuel during the core helium-burning phase means that more time is spent in this phase. Therefore, the maximum surface temperature reached during this phase is also increased. The uncertainty introduced in the specification of a blue limit is as great as that introduced by the uncertainty in nuclear burning rates (Robertson, 1972).

Uncertainties in envelope opacity and in the extent of overshooting at the base of the convective envelope influence the location of the red limit of the helium-burning band (shown in Figure 1 but not in Figure 3). This red limit is related to the appearance of convection in the envelope.

A final uncertainty in theoretical models is the brightness of crossings of the instability strip following the core helium-burning phase. For the input physics used to construct models whose tracks are shown in Figure 3, these additional crossings are at roughly the same brightness as the first two major crossings. This is not true for other models in the literature (e.g. Kippenhahn *et al.*, 1966; Hofmeister, 1967). Uncertainties in opacity, in nuclear cross sections, and in the extent of mixing during core helium-burning will affect the details of the multiple crossings of the instability strip.

(ii) Among the uncertainties that affect pulsation results, I will here mention only those which affect the determination of blue edges for pulsation in the fundamental and first harmonic radial modes. These blue edges are best calculated in the linear, non-adiabatic approximation. Discussion can focus on the information in Figures 4 and 5 (from Iben and Tuggle, 1972a, b).

No satisfactory way of treating convection in a dynamic situation has been devised (not surprising, since neither has a treatment of convection under steady state conditions been devised). However, assuming that at any point the flux of energy in convective motions remains constant during pulsation and of the same magnitude as in the static initial model, some progress can be made in estimating the influence of convection (Baker and Kippenhahn, 1965). Adopting a mixing length treatment with mixing length given by a local pressure scale height, one finds that the surface temperature at a blue edge can increase by as much as $\Delta \log T_e \sim 0.02$–$0.03$ relative to the situation when convection is completely neglected. However, with this treatment of convection, the convective flux in some regions exceeds a theoretical maximum (Christy, 1966). Insisting that convective flux cannot exceed the theoretical maximum reduces the shift to $\Delta \log T_e \sim 0.005$ (Iben and Tuggle, 1972a).

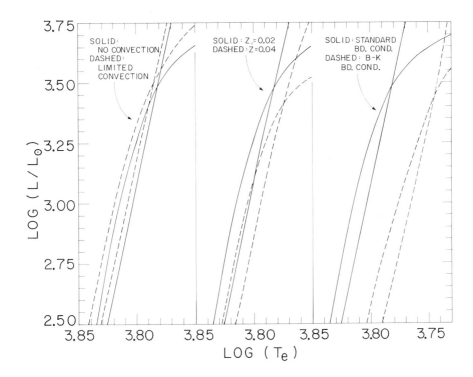

Fig. 4. The dependence of blue-edge location on convection, opacity parameter Z, and surface boundary conditions. In each panel, the solid lines are blue edges for pulsation in the fundamental and first harmonic modes for a $5 M_\odot$ model of composition $X=0.7$, $Z=0.02$. The dashed lines are blue edges for models that differ from the standard set in one characteristic: (a) convection is included, (b) $Z=0.04$, (c) Baker-Kippenhahn surface boundary conditions are used. Lines with the least curvature are fundamental blue edges (from Iben and Tuggle, 1972a).

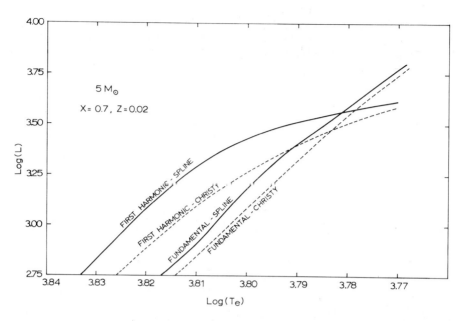

Fig. 5. Blue edges for pulsation in the fundamental and first harmonic modes. Stellar mass is $5M_\odot$ and composition parameters are $(X, Z) = (0.7, 0.02)$. Dashed lines are obtained by using Christy's (1966) analytic approximation to Cox and Stewart (1963) opacities. Solid lines are obtained by using cubic spline interpolation in fine-grain Cox and Stewart (1972) opacities (from Iben and Tuggle, 1972b).

The location of a theoretical blue edge is affected by the choice of surface boundary conditions, but the uncertainty thereby introduced is now probably negligible. This was not always true. When boundary conditions are incorrectly applied at the photosphere rather than at very small optical depth, blue edges can shift as much as $\Delta \log T_e \sim$ ~ 0.03–0.04 to the red (see Figure 4 and Iben and Tuggle (1972a)).

The choice of opacity has a major influence on the location of a calculated blue edge. This is not surprising since it is the nature of the opacity in the helium and hydrogen ionization zones that is in large part responsible for driving pulsation. What may be surprising is that even two very good approximations to the same set of opacity tables may give quite different results, as is demonstrated in Figure 5 (from Iben and Tuggle, 1972b).

(iii) The uncertainties attending an interpretation of observational characteristics of a star in terms of characteristics that may be directly compared with theory have already been alluded to in Section 2. One of the gravest uncertainties lies, of course, in the estimation of absolute magnitude. The so-called 'photometric' method works only for a dozen or so Cepheids in the Galaxy that happen to occur in young clusters with appreciable main sequence populations. One assumes that the distance to the relevant cluster can be estimated by fitting this main sequence to a 'standard' main sequence. One basic weakness of the method is the assumption that composition differences may be ignored both in constructing the standard and in fitting to the standard. There is the further difficulty of determining the distance to the Hyades cluster. This latter distance, of course, determines the normalization of the standard. One might expect an *a priori* uncertainty of 0.25–0.5 mag. in the bolometric magnitude of any of the dozen Cepheids estimated by the photometric method.

The current method for estimating magnitudes for Cepheids not in clusters rests very heavily on the slope of a mean relationship between period and color determined for galactic Cepheids and on the slope of a mean relationship between period and relative magnitude for Cepheids in systems external to our Galaxy (e.g., Sandage and Tammann, 1969). This latter relationship is normalized by means of the dozen Cepheids whose magnitudes have been estimated photometrically. Crude pulsation theory and estimates of a mean relationship between color and surface temperature are used to establish the form of a relationship between magnitude, color, and period. The coefficients in this final relationship are obtained by using the period-color and period-magnitude relationships. The most severe weakness of this PLC (period-luminosity-color) method of estimating magnitude is the difficulty in estimating a mean period-color relationship. The scatter of the observed distribution in the period-color plane is so large that only by luck will one choose a slope for the mean relation that is consistent with pulsation theory. Given $B-V$ and P (see Iben and Tuggle, 1972b), errors as large as ± 0.5 mag. in estimates of luminosity may be expected.

Another characteristic one wishes to obtain from the observations is surface temperature. Unfortunately theoreticians have not yet presented us with composition-dependent conversions between the observational means, $\langle B-V \rangle$ and $\langle B \rangle - \langle V \rangle$, and the surface temperature T_e that would characterize the star if its pulsational amplitude were vanishingly small. However, the necessary conversion formula, even if available, would be very complex and require accurate specification of additional quantities that are very difficult to obtain for distant Cepheids: the degree of microturbulence and the abundance of the major contributors to opacity (see Bell and Parsons, 1972).

3.3. Comparison between Observation and Theory

Considerable attention has been paid over the last three years to the fact that each of four different ways of estimating Cepheid masses gives a different result (Cogan, 1970; Rodgers, 1970; Fricke *et al.*, 1972; Iben and Tuggle, 1972a, b). I will discuss only two ways of estimating mass and argue that the apparent discordance is due simply to an underestimate of Cepheid luminosities (Iben and Tuggle, 1972a).

The first method is based on evolutionary calculations. For any choice of composition, evolutionary calculations establish a mean relationship between mass and luminosity during the core helium-burning phase. When $X=0.7$, $Z=0.02$, this relationship is approximately $\log M \sim 0.726 + 0.251 \times (\log L - 3.25)$. For any mass, the dispersion in luminosity about the mean is only ± 0.25 mag. For any luminosity, the dispersion in mass is only $\Delta \log M \sim \pm 0.025$. There are three studies in the literature that permit one to tentatively estimate the composition dependence of the normalization of the evolutionary $M-L$ relationship (Hallgren and Cox, 1970, Robertson, 1971; Noels and Gabriel, 1974). They suggest that for stars of mass $\sim 5\,M_\odot$, $\log M \sim 0.726 + 0.251 \times (\log L - 3.25) + (X - 0.7) + 3(Z - 0.02)$. Not enough information is available to determine whether or not the slope of the relationship (coefficient of $\log L$) depends strongly on the composition. However, it is clear that, given an estimate of luminosity and composition, one may in principle estimate a Cepheid's 'evolutionary' mass M_{evo}.

The second method is based on linear pulsation calculations which establish a relatively composition-independent relationship between mass M, fundamental period P_F, luminosity L, and surface temperature T_e. One approximation to this relationship (Iben and Tuggle, 1972a) is

$$\log M \sim 1.7 - 1.5 \log P_F + 1.26(\log L - 3.25) - 5.25(\log T_e - 3.77).$$

Note that the dependence on P, L, and T_e given by this relationship,

$$M_{\text{puls}} \tilde{\propto} L^{5/4} T_e^{-5.25} P^{-3/2},$$

is significantly different from the relationship,

$$M_{\text{puls}} \tilde{\propto} L^{3/2} T_e^{-6} P^{-2},$$

so frequently used in earlier days to help devise a PLC relationship.

Pulsation masses for 13 galactic Cepheids for which luminosities can be estimated photometrically are shown in Figure 6. At any given luminosity, the mass suggested by evolutionary calculations for the composition $X=0.7$, $Z=0.02$ is larger by about 40% than the pulsation mass.

The difference between the two estimates of mass can be minimized in several ways: (a) by increasing the helium abundance used in evolutionary calculations by $\Delta Y \sim 0.15$; (b) by decreasing the surface temperature assigned to all Cepheids by about $\Delta \log T_e = 0.025$, or (c) by increasing the luminosity assigned to all Cepheids by about $\Delta \log L \sim 0.1$ (0.25 mag.). The first two alternatives can be discarded by examining the situation in the HR diagram (Figure 7). Each of these alternatives leaves too large a gap between fundamental mode blue edges (for the appropriate masses) and the location of the 13 Cepheids. If this argument is not convincing, one may argue that a helium abundance $Y \sim 0.45$ is far larger than Y estimated for other galactic objects and that an error of 0.025 in the conversion from $B-V$ to $\log T_e$ is not to be expected. The final alternative requires that the normalization of the standard main sequence is incorrect by about 0.25 mag. It is interesting that every method of estimating the distance to the Hyades, other than the convergent point method (which is now used to set the standard), makes the Hyades distance modulus greater by about 0.2 mag. than the one given by the convergent point method (van Altena, 1973). Thus, it appears very likely that the apparent discrepancy between evolution and pulsation masses is simply due to an underestimate of Cepheid luminosities.

If, then, mass loss does not occur to an appreciable extent prior to or during the Cepheid stage, one may combine the results of evolutionary calculations with the results of pulsation calculations to derive a theoretical PLC relationship that should

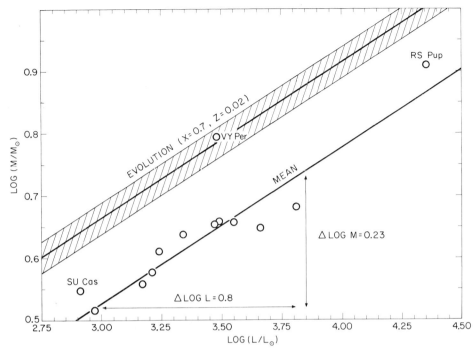

Fig. 6. Mass-luminosity relation for Cepheids given by evolutionary models ($X=0.7$, $Z=0.02$) without mass loss compared with one obtained by applying results of pulsation theory to a current interpretation of the observed properties of Cepheids in galactic clusters and associations. The shaded region is the evolutionary M-L relationship and the line labeled 'mean' is a fair representation of the pulsation M-L relation. Cepheid luminosities are tied to an assumed distance modulus of 3.05 mag. for the Hyades.
(Iben and Tuggle, 1972a.)

be an even more reliable indicator of Cepheid luminosities (and hence distances) than is the PLC relationship one attempts to derive 'from observations alone.' For the composition $X=0.7$, $Z=0.02$, the result is (Iben and Tuggle, 1972b):

$$M_{\text{Bol}} \sim -0.96 - 3.76 \log P_{\text{F}} - 13.0 \left(\log T_{\text{e}} - 3.77\right).$$

Using the transformations $M_{\text{Bol}} \sim M_V + 0.145 - 0.332(B-V)$ and $\log T_{\text{e}} \sim 3.886 - 0.175 (B-V)$ which are thought appropriate for galactic Cepheids, the theoretical PLC relationship transforms into

$$M_V \sim -2.61 - 3.76 \log P_{\text{F}} + 2.60 (B-V).$$

Neither this PLC relationship nor any other can be applied to all Cepheids without thinking. Above all, the effects of composition differences must be properly included. This means that sufficient evolutionary calculations must be done to determine, as a function of composition, the slope as well as the normalization of the mass-luminosity relationship. It means further that sufficient model atmosphere calculations must be done to determine the manner in which the conversion between T_{e} and $B-V$ depends on the composition (Bell and Parsons, 1972).

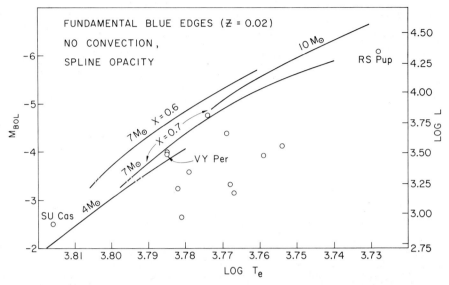

Fig. 7. Fundamental blue edges in the HR diagram for masses $M/M_\odot = 4, 7, 10$ and composition ($X = 0.7$, 0.6; $Z = 0.02$; no convection) compared with the estimated surface temperatures and magnitudes of Cepheids in galactic open clusters and associations. Cubic spline interpolation in Cox and Stewart (1972) tables has been employed to obtain the opacity. The blue edge of $X = 0.6$ is an estimate based on a linear extrapolation of blue edges for $X = 0.7$ and 0.8. The introduction of convection shifts each blue edge to the blue by approximately $\Delta \log T_e \sim 0.005$. Data from Tuggle and Iben (1974).

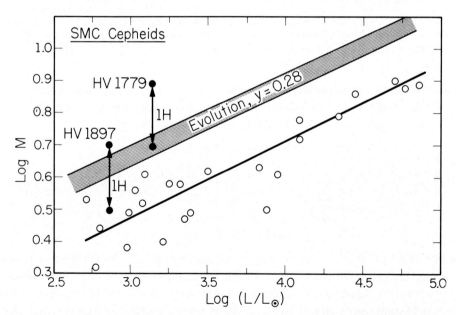

Fig. 8. Mass-luminosity relationship for Cepheids given by evolutionary models compared with a mass-luminosity relationship defined by Cepheids in the Small Magellanic Cloud if (1) the distance modulus for the SMC is 19.25 mag. and (2) if the $(B-V)$ to $\log T_e$ conversion is identical to that used for galactic Cepheids. Data from Gascoigne (1969); Theoretical estimates from Tuggle and Iben (1974).

To show how important these requirements are, let us look at estimates of pulsation and evolution masses for Cepheids in the Small Magellanic Cloud (Gascoigne, 1969). If luminosities are determined by adopting ('arbitrarily') a distance modulus of 19.25 mag. and if surface temperatures are determined by using the color-to-temperature conversion thought appropriate for galactic Cepheids (Kraft, 1961), the pulsation masses shown in Figure 8 result (Tuggle and Iben, 1974). On comparing with evolu-

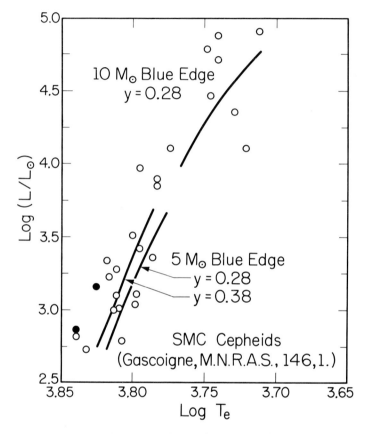

Fig. 9. Fundamental blue edges in the HR diagram for masses $M/M_\odot = 5$ and 10 and composition ($X = 0.7, 0.6, Z = 0.02$; no convection) compared with estimates of surface temperature and luminosity for Cepheids in the Small Magellanic Cloud. Luminosities (Gascoigne, 1969) are based on a distance modulus of 19.25 mag. and temperatures are based on a conversion thought appropriate for Cepheids in our own Galaxy. Theoretical data are from Tuggle and Iben (1974).

tionary masses, the mass discrepancy again appears. However, a more serious discrepancy appears in the HR diagram (see Figure 9). The Cepheids are all much bluer than the blue edges for the relevant masses. The way out of both difficulties is to argue that the average 'metallicity' of SMC Cepheids is much smaller than that of galactic Cepheids and that the conversion from $B-V$ to $\log T_e$ that is appropriate for galactic Cepheids gives too large a surface temperature for the SMC Cepheids (Bell and

Parsons, 1972). If one arbitrarily moves all Cepheids in Figure 9 to the red by $\Delta \log T_e \sim 0.02$, then the discrepancy in the HR diagram is removed. And, wonder of wonders, the discrepancy between pulsation and evolution masses is also removed. Thus, by a lucky accident, the distance modulus remains unchanged from the original choice, for whatever reasons that choice was made.

4. Population II Variables

4.1. Gross differences between population I and population II variables

The distinction between the type of variable found in globular clusters and the type of variable found predominantly in the galactic disk is most strikingly displayed by

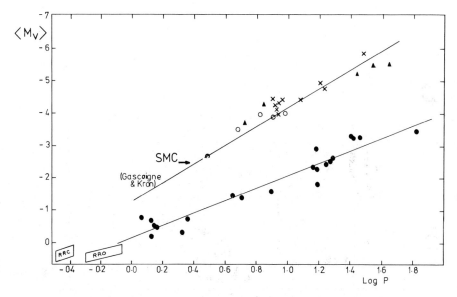

Fig. 10. Approximate magnitude-period relationships defined by variables belonging to two distinct populations. The upper curve is for population I and the lower curve is for population II. This figure is reproduced from Dickens and Carey (1967).

the difference in the magnitude-period relationships appropriate to the two types. As shown in Figure 10 (from Dickens and Carey, 1967), the population I variables are significantly brighter at any period than are the population II variables. At a period of 10 days, the difference of about 2 mag. is most easily accounted for if masses differ by approximately a factor of ten, the population I variables of this period having a mass near 5–7 $M_\odot$.

The second most striking difference between the two types is in the slope of the magnitude-period relationship, the slope for population I variables ($-dM_V/d \log P \sim \sim 3$) being much larger than the slope for population II variables ($-dM_V/d \log P \sim \sim 1.9$). This difference is most easily accounted for if the mass of a population I variable is a monotonically increasing function of increasing luminosity and if the masses of all population II variables are roughly the same. Thus, simply by analyzing the appropriate magnitude-period relationships, one may discover that (a) the most evolved stars in globular clusters are of low mass and hence are representative of an old population, and that (b) the most evolved stars in the disk are of all different masses and hence are of all different ages.

4.2. THEORETICAL DISTRIBUTION IN THE HR DIAGRAM

We are all familiar with the observed color-magnitude distributions for globular cluster stars. Theory produces distributions that bear a reasonably close qualitative resemblance to the observed distributions. In Figure 11 is shown a typical theoretical distribution. Data from Simoda and Iben (1970), Iben and Rood (1970), Strom *et al.* (1970), Schwarzschild and Härm (1970), and Iben and Huchra (1971) has been used to construct this figure. All detail of individual tracks is suppressed in order to indicate where stars spend most of their nuclear-burning lives. A more realistic description of the distribution on the horizontal branch is given by Rood (1973).

Most of a star's lifetime (10^{10} yr for an 0.8 $M_\odot$ star) is of course spent on the main sequence where hydrogen is converted into helium at the stellar center. Appreciable time ($\sim 10^9$ yr) is also spent on the subgiant branch where hydrogen is converted into helium in a thick shell surrounding an inert helium core within which electrons are becoming degenerate. Comparable times ($\sim 10^8$ yr) are spent on the giant branch (hydrogen burning in a thin shell) and on the horizontal branch (helium burning at the center and hydrogen burning in a shell). As a star reaches the tip of the first giant branch, helium burning begins in an electron degenerate core of mass ~ 0.45 $M_\odot$. During the ensuing 'helium flash', a star jumps over to a position on the horizontal branch. The more mass it loses from the surface during the giant branch phase, the bluer the average position it adopts on the horizontal branch.

Following the exhaustion of helium at the center, a star rapidly moves to a position on the supra-horizontal branch or on the asymptotic branch, the mean position again being determined by the star's mass. There, hydrogen and helium burn in separate shells surrounding an inert carbon-oxygen core.

All stars eventually reach the asymptotic giant branch, which may extend to luminosities considerably brighter than is indicated in Figure 11. At some point high enough along the asymptotic giant branch, a thermal instability sets in. If it is sufficiently light, the star then loops back from the asymptotic giant branch in a series of relaxation oscillations (Schwarzschild and Härm, 1970). During this last phase, considerable mass may be lost from the star and it is possible that, as a consequence, the asymptotic branch may bend over toward cooler temperatures at luminosities greater than that at the tip of the giant branch proper (see Feast, 1974). It is also probable (on the basis of the theoretical calculations alone) that Mira-type pulsations of increasing amplitude may occur here and that a final large amplitude pulse may blow off all but a minute portion of the matter lying above the hydrogen burning shell (see Wood, 1974). This phenomenon may be the origin of many planetary nebulae – objects consisting of a shell of outwardly expanding matter surrounding a central star that, after burning brightly as a blue star for 10^4 yr or so, cools off to become a white dwarf.

Stars occurring in regions redward of the pulsational blue edges indicated in Figure 11 are expected to be unstable to radial pulsation in the indicated modes. Thus, one expects three groups of variables to be particularly conspicuous: (a) stars on the horizontal branch that pulsate in either the first harmonic mode or in the fundamental mode (c-type and ab-type RR Lyrae stars); (b) stars on the suprahorizontal branch

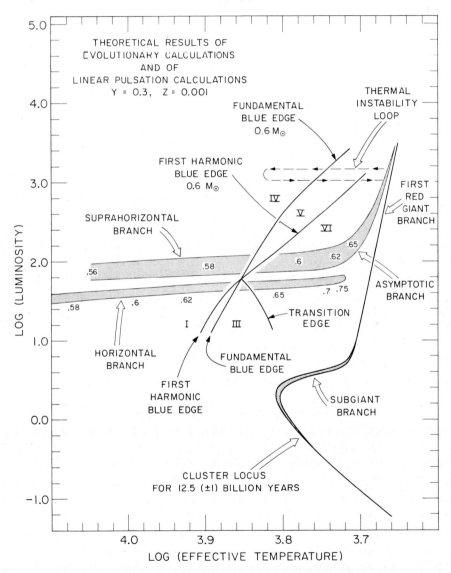

Fig. 11. Theoretical cluster locus compared with blue edges and a conjectured transition edge for $0.6 M_\odot$. Each number along the horizontal and suprahorizontal branches indicates the mass of a model whose mean location is correlated with the position of the number (from Iben and Huchra, 1971).

that pulsate primarily in the fundamental mode (BL-Herc stars or short period population II Cepheids); and (c) stars in the process of undergoing relaxation oscillations that cause them to swing back and forth from the asymptotic branch, pulsating in the fundamental mode while in the instability strip (W-Virginis stars or long period population II Cepheids).

Some additional detail should be mentioned. For sufficiently light stars, evolution during the double nuclear-shell-source stage occurs above the supra-horizontal branch, far to the blue of the asymptotic branch. Relaxation oscillations in such stars lead to wide amplitude excursions in the HR diagram that occupy a region far to the blue of the blue edges shown in Figure 11.

4.3. THE DRIVING REGIONS FOR PULSATION

In all variables that lie in the classical instability strip, the properties of hydrogen and helium ionization zones are responsible for instability against pulsation. In RR Lyrae stars with large pulsation amplitudes, these zones lie very near the surface in both mass and spacial displacement, as is illustrated in Figures 12 and 13. Figure 12 (from Iben, 1971b) describes the distribution of state variables and mass as a function of distance

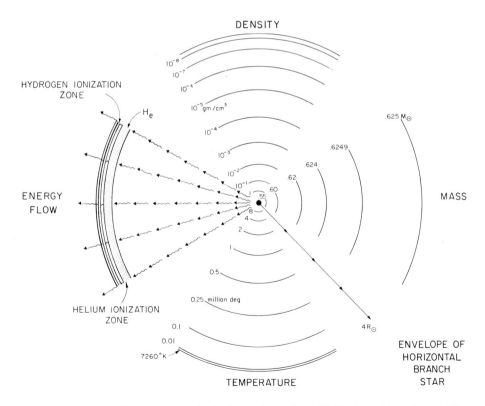

Fig. 12. Distribution of structure variables in the envelope of an initial horizontal branch model (from Iben 1971b).

from the center within the envelope of a model horizontal branch star. The total mass above the second helium ionization zone is only $10^{-7} M_\odot$ and the total mass above the hydrogen ionization zone is only $10^{-9} M_\odot$. Figure 13 (from Iben, 1971a) shows that the only positive contributions to driving pulsation come from matter in the two major ionization zones.

Although in the illustration the second helium ionization zone is the major contributor to pulsation, it is clear that the presence of the hydrogen ionization zone plays

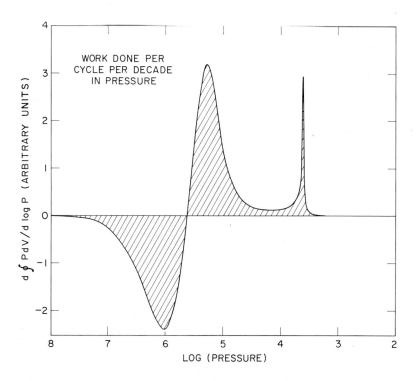

Fig. 13. Work done per cycle per decade in pressure for the first harmonic mode of the model described in Figure 12 (from Iben, 1971a).

an essential role. Certainly, the precise location of blue edges for pulsation in various modes and the phase relation between light and radius amplitudes are strongly influenced by the properties of the hydrogen ionization zone (Castor, 1971).

4.4. BLUE EDGES FOR PULSATION IN THE FUNDAMENTAL AND FIRST HARMONIC MODES

For masses and compositions thought appropriate for horizontal branch stars in globular clusters, blue edges for pulsation in the fundamental and first harmonic modes intersect each other at luminosities in the neighborhood of those defined by horizontal branch stars. At luminosities below the intersection point, the blue edge for pulsation in the first harmonic mode is at a higher surface temperature than is the blue edge for pulsation in the fundamental mode. At luminosities above the intersection point, the reverse is true. This is illustrated in Figure 14 (from Tuggle and Iben, 1972).

Also demonstrated in Figure 14 is the sensitivity of blue edges and their inter-

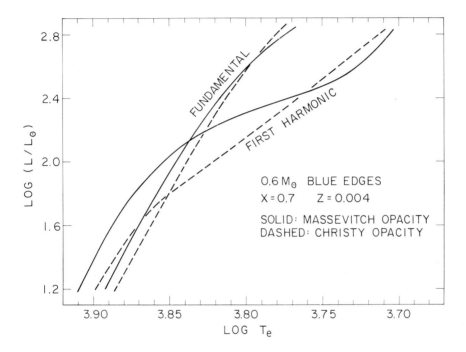

Fig. 14. Blue edges for pulsation in both the fundamental mode and in the first harmonic mode for a model star of mass $M = 0.6\ M_\odot$ and composition parameters $X = 0.7$, $Z = 0.004$. Solid curves are blue edges constructed with spline opacities (from Tuggle and Iben, 1972).

section points to the precise form of the opacity law used in making the calculations. Small differences in opacity approximations (see Figure 15, from Tuggle and Iben (1972)) are responsible for rather dramatic changes in the location of blue edges.

In all clusters that have been carefully examined and in which the RR Lyrae stars form a fairly broad distribution in color, the bluest RR Lyrae stars are pulsating in the first harmonic mode (Bailey c-type variables) whereas the reddest are pulsating in the fundamental mode (Bailey ab-type variables). This means, of course, that the intersection between blue edges must occur in the HR diagram above the horizontal branch. Since, for a given choice of opacity, the intersection point decreases in luminosity with decreasing model mass and also with decreasing envelope helium abundance, the observed relative location in color of c- and ab-type variables may be used in conjunction with observation-related estimates of horizontal branch luminosity to place lower limits either on mass (if the helium abundance is known) or on helium abundance (if the mass is known).

Estimates of helium abundance may be made in several ways. In principle, one of the

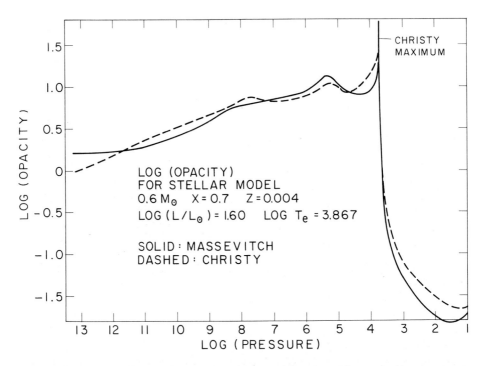

Fig. 15. Opacity derived by use of spline interpolation in fine grain opacity tables (Cox and Stewart, 1972) compared with opacity given by Christy's analytic approximation to Cox and Steward opacity tables. Opacities are plotted as functions of $\log P$, where P is the pressure (dyne cm^{-2}) in a model stellar envelope characterized by $M = 0.6 M_\odot$, $\log(L/L_\odot) = 1.6$, $X = 0.7$, $Z = 0.004$, $Y = 0.296$. This envelope has been constructed with the 'spline opacity' (from Tuggle and Iben, 1972).

cleanest ways makes use of the relationship between period P and surface temperature T_e along first harmonic blue edges. As is illustrated in Figure 16 (from Tuggle and Iben, 1972). this relationship is relatively insensitive to mass and is primarily a function of the helium abundance Y. Given that, in a particular cluster, the periods of the shortest period c-type variables are in the range $\log P \sim -0.6 \rightarrow -0.5$ and that an estimate of the surface temperature of the bluest variables is in the range $\log T_e \sim 3.86 \rightarrow 3.87$, one may estimate a helium abundance in the range $Y \sim 0.20 \rightarrow 0.25$ (Tuggle and Iben, 1972; Cox et al., 1973).

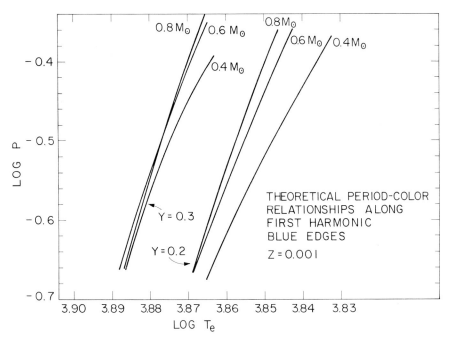

Fig. 16. The relationship between period and surface temperature for stars along first-harmonic blue edges in the major region of relevance to RR Lyrae stars in globular clusters. Curves for $M/M_\odot = 0.4, 0.6, 0.8$, $Z = 0.001$, and $Y = 0.3$ and 0.2 are shown (from Tuggle and Iben, 1972).

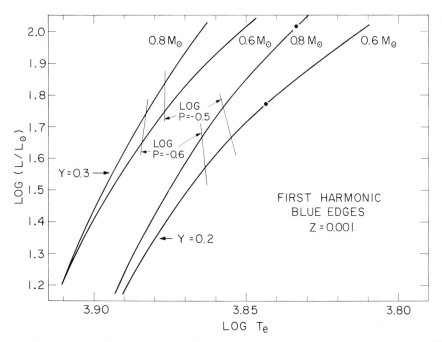

Fig. 17. First-harmonic blue edges for $M/M_\odot = 0.6, 0.8$; $Y = 0.2, 0.3$ (from Tuggle and Iben, 1972).

Once helium abundance has been estimated, one may then use additional properties along blue edges to determine a relationship between mass and luminosity, as is made clear in Figure 17. Even a rough estimate of mass permits a relatively tight estimate of luminosity. Or, given a rough estimate of luminosity, a rough estimate of mass can be derived.

Other estimates of mass and luminosity can be obtained by combining additional theoretical relationships with observational data. Evolution theory permits one to estimate both L and M versus T_e for horizontal branch stars as a function of Y and Z. Estimates of Z can of course be provided in principle by spectroscopy and/or photometry. Thus, with estimates of T_e for variable stars, one may determine both L and M as functions of Y. The composition-independent relationship between P, L, T_e, and M given by pulsation theory may be used to give a further relationship between L and M for variables. Combining the information from all approaches one may again obtain estimates of all bulk properties (see, for example, Baker, 1965, van Albada and Baker, 1971; Iben, 1971b).

4.5. THE TRANSITION REGION

One of the most interesting problems in pulsation theory that is yet to be resolved is the full-amplitude behavior of stars in the region where models are unstable to radial pulsation both in the fundamental mode and in the first harmonic mode (Christy, 1966). The current situation may be described in terms of the curves in Figure 18 (from Iben, 1971b). Brute-force calculations show that the region where both modes can be excited may be partitioned into three groups according as (a) final, full-amplitude motion is in the first harmonic mode, regardless of initial conditions (region 1); (b) final motion is in the fundamental mode, regardless of initial conditions (region 2); and (c) final motion, after a finite number of cycles, depends on the initial mode of excitation (region 3).

This latter region might be called the 'transition region' between pulsation in the first harmonic and pulsation in the fundamental mode. Its size and shape are difficult to determine by standard brute-force calculations and one might expect that conclusions could depend on the number of cycles for which the motion is followed.

The brute-force calculations suggest that, within region 3, pulsation begun in the fundamental mode will continue in the fundamental mode with no indication of switching to the first harmonic. Similarly, pulsation begun in the first harmonic mode persists with no sign of switching to the fundamental mode. Thus, a kind of 'hysteresis' occurs for models in region 3. But, is this 'type-3' behavior independent of the number of cycles for which motion is followed? Or will switching to a truly favored mode occur after a sufficiently long time?

Recent calculations by von Sengbush (1973) and by Stellingwerf (1973) may help provide an answer. Using a technique developed by Baker and von Sengbush (1969), they are able to construct the final, full-amplitude model pulsating in a pure mode and test this model for stability. Preliminary results by von Sengbush (1973) suggest that there is a finite region 3 in the HR diagram where true hysteresis occurs. However, the

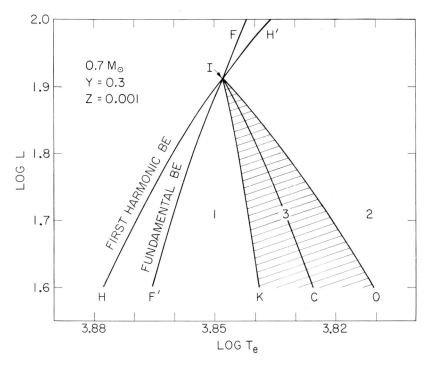

Fig. 18. Schematic description of the relative locations of different types of instability. Above point *I* and to the red of the fundamental blue edge, pulsation at full amplitude is in the fundamental mode. Below point *I*, the situation is more complex. Between the first harmonic blue edge and the curve *IK*, full amplitude motion is in the first harmonic mode, regardless of initial conditions. The region between the fundamental blue edge and the curve *IK* has been christened by Christy the region of type 1 instability. Beyond the curve *IO*, full amplitude motion is in the fundamental mode, regardless of initial conditions (Christy type 2 instability region). In the region between *IK* and *IO*, pulsation may persist in the mode initially excited (Christy type 3 instability region). The angle *KIO* may well depend on the number of cycles for which motion can be practically followed (from Iben, 1971b).

site of this region is much smaller than was suggested by the earlier brute-force calculations (Christy, 1966).

If horizontal branch stars evolve in both directions in the HR diagram the hysteresis effect should show up in the form of overlap in color between c-type (first harmonic) and ab-type (fundamental) variables. Or, if no overlap occurs, one might deduce that horizontal branch stars all evolve predominantly is one direction only.

The possible existence of a hysteresis effect permits one to account for the fact that clusters with many RR Lyrae stars fall clearly into one of two Oosterhoff (1939) types. Suppose that the mean luminosity of all horizontal branch stars is roughly the same, independent of composition, but that the direction of evolution along the horizontal branch is critically dependent on composition. In clusters in which evolution along the horizontal branch is predominantly to the right (high T_e to low T_e), stars which begin pulsating in the first harmonic in region 1 will continue to pulsate in the first harmonic as they evolve through region 3. Only when they reach region 2, will they pulsate in the fundamental. Thus there will be a high percentage of c-type variables and the 'transition period' will be large compared to the case where stars evolve predominantly from low T_e to high T_e. In this later case, stars which begin pulsating in the fundamental mode in region 2 will continue to pulsate in the fundamental mode as they evolve through region 3, switching into the first harmonic only on evolving into region 1.

It may well be, as van Albada and Baker (1971, 1973) suggest, that hysteresis plus direction of evolution are the dominant contributors to the Oosterhoff effect. However, there are other factors that almost certainly make a contribution. The additional factors include differences between clusters of the two types: (a) in the mean mass of RR Lyrae stars (Stobie, 1971; Castellani et al., 1973), (b) in the mean luminosity of the horizontal branch (e.g., Iben, 1971b), and (c) in the distribution in number vs color on the horizontal branch (Iben, 1971b).

That the distribution in color along the horizontal branch plays a role in determining the ratio of c-type to ab-type variables is made indisputably clear in Figures 19 and 20

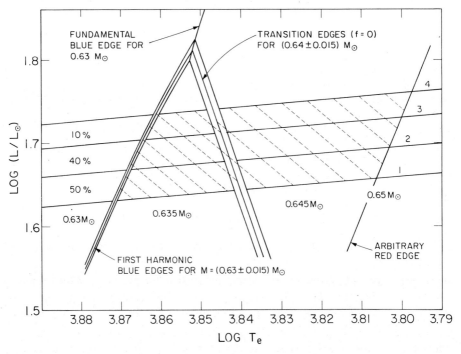

Fig. 19. Theoretical blue edges, constant period lines, transition edges, and the theoretical horizontal branch for $Y=0.3$, $Z=0.001$. Mean mass is indicated as a function of $\log T_e$. In the truncated pyramid bounded by the first harmonic blue edges and the adopted transition edges, stars are assumed to pulsate at full amplitude in the first harmonic mode. Between the transition line and the semi-empirical red edge, stars are assumed to pulsate at full amplitude in the fundamental mode. The mean position of a horizontal branch star is near line number 2, approximately one third of the way up from the base of the horizontal branch and two thirds of the way down from the top. Lines (dashed) of constant period have the slope $d \log L/d \log T_e \sim 4.04$ (from Iben and Huchra, 1971).

(from Iben and Huchra, 1971). The theoretical distribution in number vs period, shown in Figure 20, follows from the location of the blue and red edges, the location of transition edges (transition *regions* of zero width), and the location of the horizontal branch, shown in Figure 19, *if the distribution in number versus log T_e along the horizontal branch is constant.* On comparing (Figure 20) the theoretical distribution with the distribution defined by variables in the cluster M3, it is clear that the *relative* locations of edges and of the horizontal branch have been specified correctly. However, the ratio of ab-type variables to the number of c-type variables in the theoretical distribution is smaller (by a factor of 2.5) than the observed ratio. The *only* way to achieve agreement is to replace the assumption of uniform density (in number vs log T_e) along the horizontal branch by an assumed distribution that has a much higher

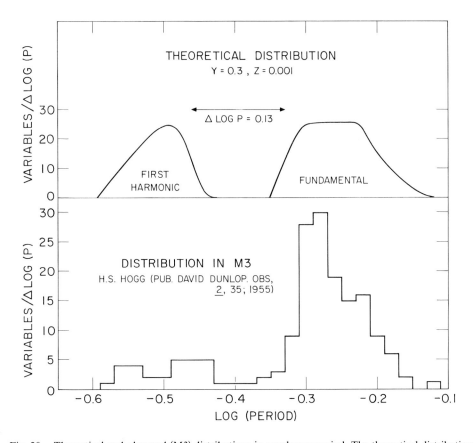

Fig. 20. Theoretical and observed (M3) distributions in number vs period. The theoretical distribution has been derived on the assumption that the density of horizontal branch stars is constant with respect to log T_e. It has been further assumed that at a given surface temperature, horizontal branch stars are distributed in the following way (refer to Figure 19): 50% between curves 1 and 2, 40% between curves 2 and 3, 10% between curves 3 and 4. In the theoretical distribution there are roughly twice as many stars in the fundamental peak as there are in the first harmonic peak. The actual ratio in M3 is approximately 5 to 1 (from Iben and Huchra, 1971).

density in the region of ab-type variables than in the region of c-type variables (see Iben and Rood, 1970, for further discussion).

One must also worry about the assumption that the dominant direction of evolution within the instability strip can alter abruptly and discontinuously when composition parameters are varied through some 'critical set.' This assumption amounts to a rather cavalier dismissal of many studies of evolution along the horizontal branch that show track behavior varying smoothly and continuously with change in composition (e.g., Iben and Rood, 1970; Sweigart and Gross, 1973).

Is it not as likely that the location of the 'transition region' changes rapidly when composition parameters vary past some 'critical set'? Certainly, the properties of the transition region are less well understood than any other element that enters into the analysis.

To this reviewer, it seems quite possible that the mean position of the transition region may move to the red (at any given magnitude and for any given mass) with decreasing abundance of the heavy elements and/or with decreasing abundance of helium in such a way as to give a much more natural accounting of the Oosterhoff effect than has yet appeared in the literature.

Acknowledgement

This research was supported in part by the United States National Science Foundation (GP-35863).

References

Albada, T. S. van and Baker, N. H.: 1971, *Astrophys. J.* **169**, 311.
Albada, T. S. van and Baker, N. H.: 1973, *Astrophys. J.* **185**, 477.
Altena, W. F. van: 1973, Invited Address to Commission Nr. 33, IAU XVth General Assembly, Sydney, Australia.
Arp, H. C.: 1967, *Astrophys. J.* **149**, 91.
Arp, H. C. and Tackeray, A. D.: 1967, *Astrophys. J.* **149**, 73.
Baade, W. and Swope, H. H.: 1965, *Astron. J.* **70**, 212.
Baker, N. H.: 1965, *Bamberger Veröff.* **4**, 122.
Baker, N. H. and Kippenhahn, R.: 1965, *Astrophys. J.* **142**, 868.
Baker, N. H. and Sengbush, K. von: 1969, *Mitt. Astron. Ges.* **27**, 162.
Bell, R. A. and Parsons, S. B.: 1972, *Astrophys. Letters* **12**, 5.
Castellani, V., Giannone, P., and Renzini, A.: 1971a, *Astrophys. Space Sci.* **10**, 340.
Castellani, V., Giannone, P., and Renzini, A.: 1971b, *Astrophys. Space Sci.* **10**, 355.
Castellani, V., Giannone, P., and Renzini, A.: 1973, in J. D. Fernie (ed.), 'Variable Stars in Globular Clusters and Related Systems', *IAU Colloq.* 21, 97. D. Reidel Publ. Co., Dordrecht, Holland.
Castor, J.: 1971, *Astrophys. J.* **166**, 109.
Christy, R. F.: 1966, *Astrophys. J.* **144**, 108.
Cogan, B. C.: 1970, *Astrophys. J.* **162**, 129.
Cox, A. N. and Stewart, J.: 1963, private communication.
Cox, A. N. and Stewart, J.: 1972, private communication.
Cox, A. N., King, D. S., and Tabor, J. E.: 1973, *Astrophys. J.* **184**, 201.
Dickens, R. J. and Carey, J. V.: 1967, *Roy. Observ. Bull.* **129**, E335.
Feast, M. W.: 1974, this volume, p. 93.
Fricke, K., Stobie, R. S., and Strittmatter, P. A.: 1971, *Monthly Notices Roy. Astron. Soc.* **154**, 23.
Fricke, K., Stobie, R. S., and Strittmatter, P. A.: 1972, *Astrophys. J.* **171**, 593.

Gascoigne, S. C. B.: 1969, *Monthly Notices Roy. Astron. Soc.* **146**, 1.
Hallgren, E. L. and Cox, J. P.: 1970, *Astrophys. J.* **162**, 933.
Hodge, P. W.: 1963, *Astrophys. J.* **137**, 1033.
Hodge, P. W. and Flower, P. J.: 1973, *Astrophys. J.* **185**, 829.
Hofmeister, E.: 1967, *Z. Astrophys.* **65**, 164.
Iben, I. Jr.: 1967a, *Science* **155**, 785.
Iben, I. Jr.: 1967b, *Ann. Rev. Astron. Astrophys.* **5**, 571.
Iben, I. Jr.: 1971a, *Astrophys. J.* **166**, 131.
Iben, I. Jr.: 1971b, *Publ. Astron. Soc. Pacific* **83**, 697.
Iben, I. Jr.: 1972, *Astrophys. J.* **178**, 433.
Iben, I. Jr. and Rood, R. T.: 1970, *Astrophys. J.* **161**, 587.
Iben, I. Jr. and Huchra, J. P.: 1971, *Astron. Astrophys.* **14**, 293.
Iben, I. Jr. and Tuggle, R. S.: 1972a, *Astrophys. J.* **173**, 135.
Iben, I. Jr. and Tuggle, R. S.: 1972b, *Astrophys. J.* **178**, 441.
Kippenhahn, R., Thomas, H. C., and Weigert, A.: 1966, *Z. Astrophys.* **64**, 373.
Kraft, R. P.: 1961, *Astrophys. J.* **134**, 616.
Noels, A. and Gabriel, M.: 1974, in preparation.
Oosterhoff, P. Th.: 1939, *Observatory* **62**, 104.
Robertson, J. W.: 1971, *Astrophys. J.* **170**, 353.
Robertson, J. W.: 1972, *Astrophys. J.* **177**, 473.
Rodgers, A. W.: 1970, *Monthly Notices Roy. Astron. Soc.* **151**, 133.
Rood, R. T.: 1973, *Astrophys. J.* **184**, 815.
Sandage, A. R. and Tamman, G. A.: 1969, *Astrophys. J.* **157**, 683.
Schwarzschild, M. and Härm, R.: 1970, *Astrophys. J.* **160**, 341.
Sengbush, K. von: 1973, *Mitt. Astron. Ges.* **32**, 228.
Simoda, M. and Iben, I. Jr.: 1970, *Astrophys. J. Suppl.* **23**, 81.
Stellingwerf, R. F.: 1973, private communication.
Stobie, R. S.: 1971, *Astrophys. J.* **168**, 381.
Strom, S. E., Strom, K. M., Rood, R. T., and Iben, I. Jr.: 1970, *Astron. Astrophys.* **8**, 243.
Sweigart, A. and Gross, P.: 1973, preprint.
Tuggle, R. S. and Iben, I. Jr.: 1972, *Astrophys. J.* **178**, 455.
Tuggle, R. S. and Iben, I. Jr.: 1973, *Astrophys. J.* **186**, 593.
Tuggle, R. S. and Iben, I. Jr.: 1974, in preparation.
Wood, P. R.: 1974, this volume, p. 101.

DISCUSSION

Rodgers: Are you saying to us that the Sandage-Tammann period-luminosity-colour relation is wrong? That is well founded on Cepheids in open clusters.

Iben: No, only that the period-colour relation is subject to uncertainty.

Rodgers: It is based on stars whose luminosities and intrinsic colours were obtained by main sequence fitting.

Iben: Yes.

Rodgers: But you say that if anyone uses the Sandage P-C relation, he is wrong and that it is the cause of the mass discrepancy in Cepheids.

I do not know of convincing evidence that η Aql is a first harmonic oscillator. There are strong reasons (e.g. beat Cepheids and Cloud Cepheids) to believe that transition periods occur around two to three days.

Iben: I have nothing against a period-luminosity-colour relationship. It is absolutely essential if one wants to estimate the bulk properties of the Cepheids in the Galaxy: their distances, luminosities and so on. You have to get it somewhere. However, what I am arguing is that one of the basic elements used in obtaining a P-L-C relationship, focussing primarily on the observations, is the P-C relation. You can get all sorts of different slopes here which will give you fantastically different results with regard to the luminosity of the given Cepheid, whereas each one of the different lines fits equally well with the observations. That is all I am saying. There is by now a lot of theory that is not all that bad that could be made use of in obtaining a P-L-C relationship. The line, given by the evolution plus pulsation theory can be passed through the relevant points to show that there is consistency with that particular observational data.

Tayler: Would it be fair to say that you were arguing, when you were comparing the evolutionary masses and the pulsation masses, that because it just looked as if you had shifted a line bodily, that you were rather unhappy with the idea of a sort of mass loss that was causing the difference. You felt you would not have got a shifted line with more or less the same slope?

Iben: It is just that if you look in the three different planes, P-L, M-L and H-R diagram, you find that you can get consistency without mass loss.

Rodgers: I think that a statement which says there is no reason to have any greater degree of mass loss rate in more massive stars than in low mass stars, is a bit sweeping because there is direct observational evidence for mass loss and where it does occur: It occurs in M supergiants and in long period Cepheids whereas it does not occur in short period Cepheids nor K giants nor ordinary giants.

Iben: I am in full agreement. It could be that this is the case. On the other hand, more massive stars spend less time as bright supergiants than do less massive stars as less bright ones. So when you fold in integration time with the lower mass loss rate, you might get the slope of that mass loss curve having exactly the opposite sign.

Tayler: The other possible argument is that if there is mass loss when they are late type stars, then you have got a very much deeper convection zone (in mass fraction) in the high mass stars than in the low mass stars. If the surface convection zone is responsible for the mass loss and if the cores of the stars do not differ very much in mass though the total masses differ a lot, then if you are going to strip something down somewhere towards the core, you would have a bigger mass loss for the high mass stars than the low mass stars.

Stobie: Are you saying that a change in slope of the $[(B-V)-\log P]$ relation actually gets rid completely of the mass discrepancy or does it change the slope?

Iben: No, it changes the slope.

Stobie: So you still need a change in the distance modulus of the Hyades.

Iben: Yes, or mass loss may be the answer.

Breger: You have a limited distribution of points. You have 13 points in this diagram and they are not randomly distributed in period, luminosity and colour, so by projecting the points into a two-dimensional diagram like this, it may not be valid to draw any slope, because part of the slope is due to the over-run from the luminosity. You have to consider all three. My question is: have you made a new solution of the data to see whether a shift like you would like is still okay with the data?

Iben: Yes, I just plonk down a line and say it looks good.

Breger: But three-dimensional?

Iben: Yes!

THE OBSERVATIONAL EVIDENCE

O. J. EGGEN

Mount Stromlo and Siding Spring Observatory, Australia

Abstract. The domain in the (U, V) plane of old-disk-population stars is defined by a sample of F- and G-type stars brighter than visual magnitude 6.5 for which luminosities are available from intermediate-band photometry. The majority of these old-disk stars avoid the domain of the young-disk population in the (U, V) plane except for (1) several objects with V near -17 km s^{-1} and (2) a half dozen possible members of a solar group. The sample of stars used to define the old-disk population is confined to distances less than 100 parsec from the Sun and those with V near -17 km s^{-1} may represent several generations of the objects formed in what is now the solar neighbourhood and which keep returning to this neighbourhood because they are isoperiodic (V near -17 km s^{-1}) with the local standard of rest. The half dozen possible members of a solar group indicate an evolutionary age of some 5×10^9 yr. In addition to some of the previously recognized stellar groups in the old-disk population (e.g. Wolf 630 and ζ Her), a new group with $U = +40$ to 50 km s^{-1} and $V = -36$ km s^{-1} is also discussed. The sample of red giants: ($V_E \leqslant 5\overset{m}{.}0$ and $(R-I)_0 > +0\overset{m}{.}42$) contains eight probable members of the halo population of which five are members of the Arcturus group. Possible structure in the old-disk-population giant sequence is discussed on the basis of published narrow-band indices, and the luminosity functions of the old- and the young-disk-population red giants are compared.

Most of the material presented can be found in Eggen, O. J.: 1973, *Publ. Astron. Soc. Pacific* **85**, 542.

DISCUSSION

Demarque: I understand that you mentioned some short period variables in M67, I never heard of such stars.

Eggen: No, all I was saying was that in the old disc population, to which M67 belongs, there are variable blue stragglers, with short periods.

Demarque: But you had some points (on your slide)...

Eggen: Those points on the slide did not all represent variable stars; only the ones that were indicated by filled circles were variables. The only cluster with a known variable amongst the blue stragglers is NGC 7789. The rest of the variables on this slide are members of groups. Also included are a few stragglers not known to be variable but which upon closer examination will undoubtedly prove to be.

Demarque: So you are suggesting that one should look at these particular blue stragglers.

Eggen: That is an obvious thing to do but the magnitudes, especially in the clusters, are rather faint.

Buscombe: Is the proper motion of R Cor Bor really reliable enough that one can talk about its kinematics?

Eggen: R Cor Bor has a long meridian history.

Rodgers: I think that it is a dangerous trap for observers to fall into, to put tags on objects which imply that we know what we are talking about, I think there is certainly point and logic in trying to say short period young disc variables in an attempt to describe β Can Maj stars. It may be a longer-winded title and it may be more explicit, but I think when one has objects like metal-rich RR Lyrae stars, which we do not see in old disc clusters at all (at least we do not see them in 47 Tucani), that kind of grey edge between the old disc and the halo which is hardly definable, then we have the old problem of more than one parameter in the colour-luminosity relation of clusters and the stellar populations including the variables. That is why I think it is a good idea to use neutral terms like W Virginis or RV Tauri and tell the theoreticians where these stars are afterwards.

Eggen: I am sure that most people do consider that W Vir stars occur in globular clusters. However, W Vir certainly is no globular cluster Cepheid. It is positioned slightly higher in the halo than the young disc stars and it has the kinematics of stars around 5 to 10×10^8 yr old. It belongs to a population which has a solar composition but it has a low mass, a feature it shares with the halo.

Iben: Can you not have a halo star in the disc?

Eggen: You can have a halo star now positioned in the disc, yes.

Rodgers: If you had a halo star which was only down by a factor of three (you said Ah!) and if metal enrichment was such that, in the genuine kinematic halo, we had stars that were only down by a factor of three, what would you say then?

Eggen: My main point is a very straightforward one. Papers are now being written that mean something different, to some readers, than what the authors intended. There may have been a time, for example, when the term RR Lyrae star meant the same thing (or perhaps nothing) to nearly everyone. Now we know that such short period variables can occur in various populations and we need some designation for this. RR Lyrae is a short period Cepheid of the halo population. AI Velorum is an ultra-short period Cepheid of the old disc population. δ Scuti is an ultra-short period Cepheid of the young disc population. To call all of these RR Lyrae stars is to cause confusion.

Feast: I have got quite a lot of sympathy with you in trying to divide things up. I think we should allow ourselves freedom to divide them up in any way we feel is suitable. What I wanted to say was even if you made this division between red variables and Cepheids, there may be quite a number of different mechanisms of instability within the strip. For instance, in the R Cor Bor stars, a star like RY Sag is pulsating as well as being an R Cor Bor star and it falls over somewhere in the Cepheid region. Yet if I understand correctly the theoretical proposal that has been made for its pulsating, the mechanism is more like a red variable mechanism although it lies in the Cepheid instability strip.

Eggen: I am saved on R Cor Bor because it is outside the Cepheid instability region but I take your point. However, there is no problem in this matter because a Cepheid is a Cepheid. But W Virginis is a Cepheid of the old disc population whereas δ Cephei is a Cepheid of the young disc population and there are known Cepheids of the halo population and the distinction is important.

Bessell: Have you found K stars in the old disc which vary?

Eggen: No, the variability in the red stars starts very suddenly at a $\log T_e$ of about 3.55 (in $R-I$, about 0".9). It is just like the blue edge on the Cepheid instability strip.

PART II

PULSATION AND THE YOUNG DISC POPULATION

RECENT PROGRESS IN LINEAR AND NON-LINEAR CALCULATIONS OF RADIAL STELLAR PULSATION*

ARTHUR N. COX

University of California, Los Alamos Scientific Laboratory, Los Alamos, N.M., U.S.A.

Abstract. This review consists of a discussion about the agreement of non-linear calculations at very low amplitude with linear theory results, the recent computational advances in progress, the current problems especially those relating to observations, a review of current computations, and finally the presentation of some recent extensive non-linear calculations.

Previous main sources of non-linear results have been Christy (1966) for RR Lyrae stars and Stobie (1969) for classical Cepheids. Now a publication by King *et al.* (1973) treats both linear and non-linear calculations and gives many new non-linear results.

Agreement between the results of linear calculations and those from a non-linear code when the pulsations are at very low amplitude is generally good, but there are some differences which hopefully are due to the necessary coarse zoning in the non-linear calculations.

The best and easiest to expect agreement between the results of linear and non-linear theories is in the pulsation periods. Extensive non-linear results have shown that non-linear period lengthening is at most 3%. Thus theoretical Q values from the period-mean density relation can be predicted highly accurately from linear theory. Therefore stellar masses derived from the relation between luminosity, effective temperature, period and mass, with the first three of these observed, will not be changed by non-linear theory results. Non-linear calculations are however useful in being sure that predicted linear pulsations can grow to observable amplitudes and that comparison with real observed stars is valid.

Blue edges (the high temperature boundary) of the pulsation instability strip for the fundamental and first harmonic modes have been shown to be closely the same in both the linear and the full amplitude non-linear theories, at least for Cepheids. The situation is confused somewhat for RR Lyrae stars because the two theories have not used identical opacity laws. Coarse zoning in the hydrogen ionization zone for non-linear calculations, results in either a slight under or overestimate of the instability driving due to the ionization of hydrogen. This very thin region, compared to zone sizes, generally causes an underestimate. However this small error can cause a stable model to be classed as pulsationally unstable even though it has a surface (or effective) temperature hotter than the hot (blue) edge from linear theory. Despite this calculational problem, no clear cut case has yet been found where there is a 'hard' self excited pulsation, i.e. where non-linear theory at a threshold amplitude gives pulsations when linear theory predicts stability. Since linear calculations are much easier and faster to make, it is desirable to know the limits of their usefulness in comparing with real stars.

---

* This work was performed under the auspices of the U.S. Atomic Energy Commission.

The phase lag of luminosity behind the time of minimum radius was originally thought to be a non-linear effect, but Castor (1968) showed that application of the proper boundary condition at the stellar surface gave the proper phase lag even in the linear theory. Non-linear effects modify the luminosity phase lag, but do not cause it.

The growth rate of the pulsation energy for all modes is the same in the linear infinitesimal amplitude theory as in the non-linear theory at very low amplitude to within 10% in the best checked case. Generally the agreement is less precise, about 50%. Pulsation driving regions are balanced against the interior damping regions and the small net driving causes amplitude growth. A slight error in the amount of driving by the hydrogen ionization zone in the coarsely zoned non-linear calculations can affect the blue edge, as discussed above, as well as influence greatly the instability growth rate.

The most stringent comparison between the theories is in the behavior of $\delta r/r$, $\delta\varrho/\varrho$, $\delta T/T$, $\delta P/P$ and $\delta L/L$ throughout the unstable star model. An example of this comparison for a 6.25-day period Cepheid is shown in Figure 1 where $\delta r/r$ vs $x = r/R$ is shown at the maximum compression stage (y_1) and then just one-quarter of a cycle

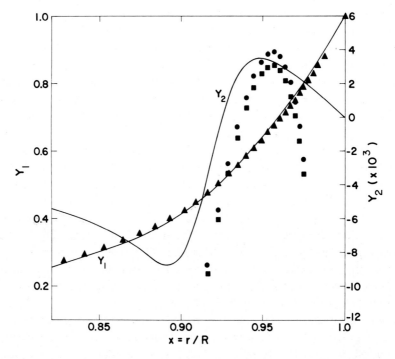

Fig. 1. The real and imaginary parts of the radius variation for a 6.25 day period 5.4 solar mass Cepheid model vs fractional radius. The y_1 linear theory result (solid line) is matched closely by the small amplitude non-linear theory results (triangles) at maximum compression. The y_2 circles and squares are at times separated by 1/5300 of a period when the outer radius is at equilibrium position moving outward, and they do not agree too well with the linear theory (solid line).

later (y_2) when the outer radius is at its equilibrium position. The 5.4 $M_\odot$ model used is identical in the linear and non-linear calculations, and has an equation of state allowing only the second ionization of helium to occur. The non-linear radius pulsation amplitude is 0.00123 and the variable y is $\delta r/0.00123\, r$. For y_2 the dots and squares give conditions in the non-linear calculation of two times separated by 1/5300 of the period centered at equilibrium radius time. While the agreement is not perfect (especially for y_2) the general behavior is similar enough to constitute a reasonable check of the non-linear programs at low amplitude.

There are several current, and possibly perennial, computational difficulties in both linear and non-linear work. Opacity variations with temperature and density need to be very accurately known in evaluating the stability of a stellar model. Until 1971, the usual opacity tables as published by Cox and Stewart (1970a, b), with their coarse temperature and density grid, were used. Christy (1966) fitted earlier data with a lengthy expression and was able to assure smooth variations with T and ϱ. However, the fit is not perfect, or perhaps it no longer fits the latest opacity data. In any case the precise location of blue edges in the HR diagram depends on whether one uses the Christy formula or more recent spline fit data.

The situation seems to be that a blue edge can be reasonably located to perhaps 0.01 or better in $\log T_e$, but the definition of the intersection luminosity of two blue edges, for example the fundamental and first harmonic blue edges, depends greatly on smooth opacity values. An accurate value of the intersection luminosity is useful in predicting possible pulsation modes.

A current calculational improvement for the radiation treatment in the outer stellar layers has been outlined by Castor (1972). Using the mean intensity as an additional variable, one can account for corrections to the radiation flow which are of the order of the material velocities divided by the speed of light. Only recent non-linear calculations by Spangenberg at Los Alamos, to be outlined later, use these latest transfer theory improvements.

Even more elaborate transport theory has been used by Keller and Mutschlecner (1970, 1971), by Bendt and Davis (1971) and by Davis (1972, 1974) for W Virginis. For Cepheids, it appears that one can predict spectral features such as colors at all phases of the pulsation period by merely constructing static atmospheres with effective temperature and effective gravity information from the radiation-diffusion models of Christy (1966) or most recently of King et al. (1973). Keller and Mutschlecner have shown for one Cepheid model that static atmospheres give color information identical with multi-frequency hydrodynamic transport atmospheres also. Detailed transport theory seems only to be needed for population II Cepheids such as W Virginis where the atmosphere is very extended and has shock waves passing through.

Another improvement being developed by Castor at Los Alamos is a mixed Lagrangian and Eulerian computation grid to define more accurately the very thin hydrogen ionization region. In linear calculations, it is possible to use enough thin zones to accurately define the hydrogen ionization region and its effects on pulsational instability. In the time-consuming non-linear work, however, zones must be thicker. In the developments underway, a few thin zones can be allowed to slide through the stellar mass to follow the hydrogen ionization detail and accurately define the driving and the light curve.

Non-linear limiting amplitudes depend on the artificial viscosity used to smooth the hydrodynamic computations. This has been studied in detail in the work of Bendt and Davis (1971). When comparison of predicted amplitudes with observations are made, especially for the problem of Cepheid pulsation modes, care must be taken to avoid the strong dependence on the artificial viscosity. Generally a rather large viscosity is used to speed the calculations, and the resulting limiting amplitude behavior is not always accurately determined. Stellingwerf at Boulder has now suggested that no artificial viscosity be used except when zone compression rates exceed a specified threshold.

The theory of energy flow due to convection is always in a somewhat uncertain state in stellar interior calculations, and if the stellar structure is continually changing this time-dependent convective flux is essentially unknown. Simple phenomenological procedures have been outlined by Cox et al. (1966) and by Cox (1967), and even used for red variables by Keeley (1970) and Wood (1973). Such methods have not been used to study stars in the usual pulsation instability strip. Indeed a discussion by Castor (1971) indicates that agreement between phase shift predictions and observations seems to require lower convective flux than the current time-independent mixing length theory gives. Thus convection is usually neglected, but discussion of red edges will eventually require a theory of time-dependent convection.

The most extensive efforts at present for the improvement of non-linear calculations are development of techniques for finding non-linear periodic solutions. Baker and von Sengbusch (1969) have treated periodic pulsations as a non-linear eigenvalue problem rather than as an initial value problem. They have used a Newton-Raphson technique to adjust the radius, velocity, internal energy, and specific volume for each Lagrangian mass point at a starting phase so that after one complete period conditions repeat exactly. The procedure involves integrations through time just as the initial value problem method, but by iterations, the time integrations need only be made for as many periods as necessary until convergence to a strictly periodic solution is found. This method has been modified by Stellingwerf (1973) to make it more adaptable to the usual initial value techniques. Further following Baker and von Sengbusch, it is possible to make a linear stability analysis of the derived non-linear periodic solution to see if it will want to switch to another pulsation mode.

There is a great advantage to this technique of finding the non-linear strictly periodic solutions, because if e-folding times are long, direct integrations for possibly thousands of periods can be avoided. Another advantage is in the analysis to see if the solution is stable or whether it tends to a neighboring solution. However, even if the periodic solution is not stable, the change to another harmonic may not actually occur. The assumed linear perturbations to the periodic non-linear solutions may not grow to observable amplitude, and the change may not completely or even partially take place.

A current problem relating to observations is to describe which modes actually are predicted at observable amplitudes. For example, observations indicate that first harmonic Cepheids and those with both fundamental and harmonic modes seem to be rare and at a rather short period (Fitch, 1970). It is known from linear theory for a given mass, that there is a maximum luminosity for which one can get harmonic pulsation. For higher luminosity only fundamental pulsations can occur. An accurate linear theory mapping of the HR diagram for harmonic pulsation can then help sort observations for the helium content in the ionization region if the mass (from the luminosity, effective temperature, period, mass relation) is assumed known.

Pulsation in the first harmonic mode occurs at lower luminosity than that at the intersection of the first harmonic and fundamental blue edges and to the red of the first harmonic blue edge. Even somewhat redward of the fundamental blue edge, first harmonic pulsations can persist until the fundamental growth rates from the linear theory increase with cooler temperatures to about $\frac{1}{8}$ that of the first harmonic pulsations. At this and cooler temperatures there is the possibility of either harmonic or fundamental pulsations (Christy type 3 behavior). At very cool temperatures, where King et al. (1973) get a fundamental growth rate of $\frac{1}{2}$ that of the first harmonic, only fundamental pulsation seems to exist. Mapping of these two transition lines which originally were thought to be almost coincident by Christy (as observations of globular cluster RR Lyrae stars seem also to show) is an important goal for non-linear studies.

For periods of Cepheids between 7 and 10 days, light and velocity curves show bumps which Christy (1968) has shown to be echoes off the hard stellar core of the previous period ionization region pressure waves. Further study by Stobie (1969) has shown that these bumps indicate low Cepheid masses, perhaps only half that given by evolutionary theory for a given luminosity. Even though these echoes seem to be correctly analyzed in terms of the mass, if luminosities and internal opacities are correct, unpublished attempts at Los Alamos to confirm this sensitivity of bump phase with mass have been unsuccessful.

A small observation-theory problem still exists in deciding how to make the proper mean of the observed color variations to plot a single position of a pulsating star on the HR diagram. A common procedure is to take the mean B intensity over time and subtract the mean V intensity over time and then use the $\langle B \rangle - \langle V \rangle$ in the Kraft (1961) relation to get $\log T_e$. Other methods are to get $\langle B-V \rangle_{mag}$ and $\langle B-V \rangle_{int}$ where these are time means of magnitudes and intensities. It is assumed that this $\log T_e$ is the same as in a static model before pulsations have grown to observable, non-linear amplitude. A comprehensive check of this assumption has not been published using non-linear calculation results as though they were observations.

Non-linear calculations now in progress seem to include only some conventional initial value integrations by Spangenberg at Los Alamos and at the University of Colorado and the search for periodic non-linear solutions by von Sengbusch at Munich and by Stellingwerf at the University of Colorado. Few results are currently available from the periodic solution searches, though checks of the method using Christy's RR Lyrae star models are well in hand.

Figure 2 gives on the theoretical HR diagram small circles for luminosities and effective temperatures of 40 models studied for pulsation by Spangenberg using non-linear techniques. Fundamental (FBE) and first harmonic (1HBE) blue edges are shown according to linear theory using the same material properties. Also shown is an estimated red edge (ERE) which is not possible to compute because of the earlier described convection problems. A line labelled PIM is the locus of points with the

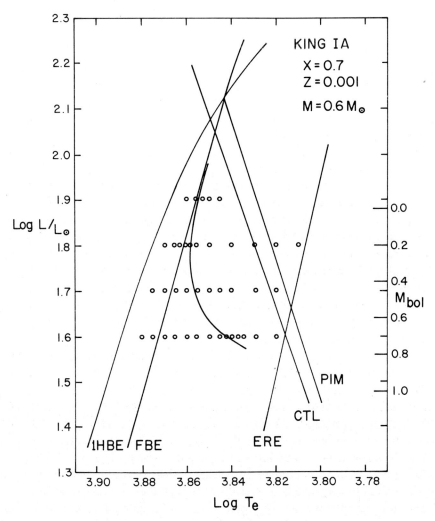

Fig. 2. The theoretical HR diagram for 0.6 solar mass models with the King Ia composition. Shown are the linear theory first harmonic blue edge (1HBE), the linear theory fundamental blue edge (FBE), the estimated red edge (ERE), The Christy transition line (CTL), the period independent of mass line (PIM), and the Spangenberg estimate of the transition line between the region of first harmonic pulsators only to pulsators with mixed harmonic and fundamental modes. The 40 models studied by the non-linear theory are given by circles for each L, T_e values.

period independent of mass and intersecting the intersection of the first harmonic and fundamental blue edges. As plotted, this line is the L, $\log T_e$, $M = 0.6\, M_\odot$ relation derived from

$$L = 4\pi R^2 \sigma T_e^4 \qquad (1)$$

$$Q = \pi \sqrt{\varrho/\varrho_0} \qquad (2)$$

and the pulsation mode transition relation

$$\log L = A \log \pi_{tr} + B. \qquad (3)$$

Iben and Huchra (1971) have considered this PIM line as maybe the same as the Christy transition line from non-linear calculations shown as CTL with $A = 1.67$ and $B = 2.07$ when solar units and days are used.

For the mixture King Ia and a mass of $0.6\, M_\odot$, these models show a curved transition line between the first harmonic oscillations and a region where both fundamental and first harmonic pulsations can exist depending upon which mode is initiated. Since integrations are carried out for only 100–1000 periods, it is entirely possible that given the very long time for deciding a pulsation mode, a real star may eventually pulsate in a different mode than indicated on this diagram. In no case however is only fundamental pulsation seen. Thus for this mass and composition the other transition line to only fundamental pulsation does not seem to exist.

Predictions are that there is a large region where Christy type 3 behavior (both fundamental and first harmonic possible) occurs. Unfortunately in most globular clusters the transition between modes is very sharp at a given color. Only in ω Centauri does this overlap exist. Thus further theoretical work in defining the transition line is necessary.

Most recent unpublished work by Stellingwerf has shown that the artificial viscosity used in the non-linear work seems to damp preferentially the fundamental mode pulsations near the transition line. Thus the exact location and curvature of this transition line is still uncertain, but it seems to be bluer at the lower luminosities than given by Spangenberg.

In spite of extensive non-linear calculations made to date, the most pressing need is for more such studies especially to predict full amplitude behavior such as transition line positions, the significance of bumps on the light and velocity curves, harmonic and double period Cepheids. With understanding of non-linear behavior, linear small amplitude predictions can be used with more precision to study pulsating star masses and compositions.

References

Baker, N. H. and Sengbusch, K. von: 1969, *Mitt. Astron. Ges.* **27**, 162.
Bendt, J. E. and Davis, C. G.: 1971, *Astrophys. J.* **169**, 333.
Castor, J. I.: 1968, *Astrophys. J.* **154**, 793.
Castor, J. I.: 1971, *Astrophys. J.* **166**, 109.
Castor, J. I.: 1972, *Astrophys. J.* **178**, 779.
Christy, R. F.: 1966, *Astrophys. J.* **144**, 108.

Christy, R. F.: 1968, *Quart. J. Roy. Astron. Soc.* **9**, 13.
Cox, A. N. and Stewart, J. N.: 1970a, *Astrophys. J. Suppl.* **19**, 243.
Cox, A. N. and Stewart, J. N.: 1970b, *Astrophys. J. Suppl.* **19**, 261.
Cox, A. N., Brownlee, R. R., and Eilers, D. D.: 1966, *Astrophys. J.* **144**, 1024.
Cox, J. P.: 1967, in R. N. Thomas (ed.), 'Cosmical Gas Dynamics' (Fifth Symposium), *IAU Symp.* **28**, 3.
Davis, C. G.: 1972, *Astrophys. J.* **172**, 419.
Davis, C. G.: 1974, *Astrophys. J.* (in press).
Fitch, W. S.: 1970, *Astrophys. J.* **161**, 669.
Iben, I. and Huchra, J.: 1971, *Astron. Astrophys.* **14**, 293.
Keeley, D. A.: 1970, *Astrophys. J.* **161**, 657.
Keller, C. F. and Mutschlecner, J. P.: 1970, *Astrophys. J.* **161**, 217.
Keller, C. F. and Mutschlecner, J. P.: 1971, *Astrophys. J.* **167**, 127.
King, D. S., Cox, J. P., Eilers, D. D., and Davey, W. H.: 1973, *Astrophys. J.* **182**, 859.
Kraft, R. P.: 1961, *Astrophys. J.* **134**, 616.
Stellingwerf, R. F.: 1973, private communication.
Stobie, R. S.: 1969, *Monthly Notices Roy. Astron. Soc.* **144**, 485.
Wood, P. R.: 1973 (3 preprints).

DISCUSSION

Iben: Does that transition line mean that stars to the left of it pulsate solely in the first harmonic and those to the right exhibit type 3 behaviour or pulsate only in the fundamental? I do not understand your statement.

Cox: According to the Christy designation, type 1 behaviour is to the left of a transition line where only harmonic pulsation is predicted by the non-linear theory. Type 3 behaviour is the one where the full amplitude mode is either harmonic or fundamental depending on which you deliberately excite at an amplitude just below full harmonic amplitude. The line drawn is between type 1 and type 3 behaviour. Spangenberg has not yet found any cases with only the fundamental occurring – the so called type 2 behaviour.

Iben: So there is a transition region that has an edge to it? There is not a transition line?

Cox: That is correct. We don't seem to find a sharp transition line between type 1 and type 2 behaviour.

Iben: It is extremely important because in that finite sized region to the right you can have either one mode or the other and there might be a hysteresis effect and so on.

Cox: Yes, but this is not a new result. The evolution direction may determine the observed mode. Christy observes this type 3 behaviour. As a matter of fact he also observes type 4 behaviour when both modes are present simultaneously. I suppose such type 4 behaviour may occur somewhere in the type 3 behaviour region, but none have been seen by Spangenberg in this work on RR Lyrae stars.

Iben: I do believe that at one time it was suggested that there was a transition line and not a transition region and that has been the whole bone of contention for a number of years. This is the first time that it has been cleared up.

Cox: Let us look again at the luminosity-effective temperature diagram. Here is the pulsation instability strip. This is the harmonic blue edge and here to the red at low enough luminosity (below the edge intersection) is the fundamental blue edge. Even redder yet is the transition edge from type 1 to type 3 behaviour. In principle, further red is type 2 (fundamental only) behaviour but this has not been found yet by Spangenberg even very close to the estimated red edge.

It appears that there is really a transition region with type 3 and type 4 behaviour possible. How this broad region fits with the sharp transition line from observations is not clear. van Albada and Baker (*Bull. A.A.S.* **3**, (1971), 241) suggest the importance of evolution direction, and this is further discussed without resolution by Iben (*Publ. Astron. Soc. Pacific* **83** (1971), 697).

Rodgers: You said earlier about the Cepheids that you were not convinced that there were any harmonic oscillators. What about the beat Cepheids? These must be in the transition zones.

Cox: What are these?

Rodgers: I call them beat Cepheids, multiple period Cepheids.

Cox: As far as I know, we have never seen Cepheids in the pure harmonic mode. We may not recognize them or they may not exist. At any rate they should be bluer and at a lower amplitude than fundamental Cepheids. Damping mechanisms limit harmonic pulsations at lower amplitude than for fundamental pulsations – just as for the c-type RR Lyrae stars.

The beat Cepheids (type 4 behaviour) are here in my diagram. If you decrease the Cepheid mass, the harmonic and fundamental edge intersection is at a lower luminosity and harmonic pulsators would occur only at low luminosity. The same luminosity decrease of the intersection point occurs if the helium content is decreased. Thus for the beat Cepheids, there must be a type 4 region at the observed low luminosity and short period obtained either by normal helium and low mass, or at lower helium content and normal evolutionary masses. I am not sure one can ever solve the mass problem of beat Cepheids because one can not know the helium content in the helium ionization region.

Irwin: From the point of view of an observer, what does the theorist need in regard to these bumps? Does he need more information and what kind of information? More detailed information? Is it important?

Cox: I am afraid the shoe is on the other foot. I think the problem now about bumps is that we just need to sit down and do these laborious non-linear calculations with great detail. I do not think we need to have more observations. There are a number of good velocity curves, a handful, maybe half a dozen or ten, which I think is good enough at the moment. Of course, there are hundreds of light curves that have the shoulder bump but are they really precise? I am not sure. My answer to you is that I do not believe we need any more observations. We need to work harder ourselves.

Iben: Did I understand you correctly to state that there may be no first harmonic Cepheids?

Cox: I only said that because I do not know, but I would really like to hear somebody who knows the answer.

Iben: Could I show some slides that show that there may be first harmonic pulsators? I think these are Small Magellanic Cloud stars and I would like to focus on these two fellows, indicated by black filled circles, relative to the others. I would contend that these are first harmonic pulsators. (See Figure 9 in Iben's paper.)

Cox: What edges are those? Fundamental blue edges?

Iben: Yes, those are fundamental blue edges, but we all know that the Small Magellanic Cloud stars are bluer because they are metal poor. If we used the correct colour temperature conversion, all these would be over here. But the main point is that these stars are amongst the bluest of a homogenous set.

Now let us look at them in the next slide and do the same sort of comparison that one does for galactic Cepheids and note that these black fellows seem to be quite far up above the line suggested by the bulk of the Cepheids in the mass-luminosity diagram. If one now assumes that these are first harmonic pulsators instead of fundamental, which was done, they come back down sort of within the range of the others. (See Figure 8 in Iben's paper.)

Cox: These may just have more helium or something. I was hoping that you would say: here is a light curve and the amplitude is low like it is for Bailey-type-c stars or it is very symmetrical instead of askew. Maybe somebody can present those light curves?

Rodgers: I have seen Gascoigne's light curves and they are symmetrical and low amplitude.

Buscombe: The theoreticians have spoken very confidently about the left edge of this instability strip. From an observational point of view, there is something going on on the upper left of the left edge which looks like semi-periodic shock waves.

Cox: I do not understand. You mean these are more blue than the blue edge?

Buscombe: Yes, certainly. Supergiants in which there is something that repeats itself approximately in, say, 7 to 30 days.

Cox: Is there a light curve or is there just spectroscopic evidence?

Buscombe: Mostly changes in the spectrum. We will talk about that tomorrow.

Stobie: In connection with the first overtone and fundamental pulsators, there is the work of the Gaposchkins on stars in the SMC. They isolated a group in the period-luminosity plane which had sinusoidal shaped light curves of lower amplitude. It turned out that, in the L-P plane, these stars lay systematically above the other stars which have larger amplitudes and more asymmetric type of light curves. They did not say anything about this but I think this is very likely that these stars, which appear at periods less than about five days, were probably first overtone pulsators. This would agree reasonably well with the transition period which we observe in our own Galaxy among the beat Cepheids which occur in the region of two to four days that you get with both modes present.

Cox: I am glad I raised this question because I have learned there are some harmonic oscillators in the Cepheids. Are there any in our Galaxy?

Rodgers: Yes, the beat Cepheids.

Zahn: Just a comment on your comparison between linear and non-linear calculations. You seem to worry about the discrepancy that you find in $Y2$ (see Figure 1). I think this is a genuine difference because it is really implied by the theory. In a non-linear calculation you have an energy source which is the κ mechanism and a sink which is radiative damping and you have an exact balance between the source and the sink. In the linear calculations, the source predominates and the excess energy is transformed into kinetic energy to build up the amplitude. So you have a different energy flow in the two cases and you must have a difference between the linear and non-linear theories.

Cox: At least there is a great attempt in the linear theory not to get the linear expression of the differential equations but to get the linear expression of the difference equations which are used in the non-linear calculations. There has been a great attempt to get the physics the same.

Zahn: What I am saying is that the eigenfunctions that you expect in the linear and non-linear theories must be different. At least the imaginary part, which is your $Y2$, has to be different, otherwise the energy cannot be conserved.

Cox: When I refer to the non-linear theory, I mean using the non-linear program at low amplitude. Thus there is neither a balance of driving and damping in either theory. Hopefully they are the same with both the linear and non-linear methods, however.

Rodgers: I think part of your question was: are there any harmonic oscillators? May I address a question to Professor Eggen? Is it not true, at least among field short period Cepheids of the order of two days, that you could not put your hand on your heart and say you know the intrinsic colours well enough to say whether it is a fundamental or first harmonic oscillator?

Eggen: Yes. There are several that could be.

Cox: If there are lots of harmonic oscillators, then that means that the transition point is at a reasonable luminosity and then probably the masses of the Cepheids are – barring the fact you can trade off helium content – probably close to the evolutionary masses.

POPULATION I CEPHEIDS – THEORY AND OBSERVATION

R. S. STOBIE

Mount Stromlo and Siding Spring Observatory, Australian National University, Canberra, Australia

Abstract. The quantitative discrepancies between theory and observation with regard to population I Cepheids are discussed. The discrepancies cannot be reconciled by mass loss or by changes in the luminosity or effective temperature of the variables. It is concluded that the present pulsation models both linear and non-linear are in error.

1. Introduction

In this paper, I will concentrate on the disagreement between theory and observation with regard to population I Cepheids. Qualitatively, there is fair agreement between theory and observation in that evolutionary tracks predict the number-period distribution of observed Cepheids, unstable pulsation models are found in the observed strip and the light and velocity curves of non-linear pulsation models compare favorably with observation even in details such as bumps in the light curve. Quantitatively, however, a comparison of theory and observation reveals a number of discrepancies for which a variety of explanations have been proposed in the literature. Each discrepancy may be expressed in different ways and we choose here to express it in terms of the mass of a Cepheid.

The mass of a population I Cepheid may be calculated by at least three methods. (a) From the Cepheids in open clusters and associations; (b) from Cepheids with a modulated light curve and (c) from Cepheids with a bump in their velocity curve. Each method is discussed in turn and finally we comment on the apparent discrepancy in the transition period from fundamental to first overtone pulsation.

In what follows, M and R denote the mass and radius of a star in solar units.

2. Calibration Cepheids

Sandage and Tammann (1969) have listed 13 Cepheids which have been used to calibrate the zero point of the period-luminosity-color relation (PLC). These 13 Cepheids are considered to have the most accurately determined absolute magnitudes and unreddened colors. This enables the radius of each Cepheid to be calculated. Given P and R, the theoretical PRM relation may be used to calculate the mass. The most complete derivation of the PRM relation for population I Cepheids is that given by Cox *et al.* (1972) but it is sufficient for our purpose to use the approximate relation for fundamental mode pulsators

$$P_0 = \alpha R^{7/4} M^{-3/4}, \tag{1}$$

where α is a constant ($\alpha = 0.022$, Christy, 1966). The mass derived in this way is called the pulsation mass, M_Q.

The mass M_Q is compared with the evolutionary mass, M_{ev}, which is derived from the luminosity of evolutionary models during the core helium-burning stage. A systematic discrepancy is found for the 13 calibrating Cepheids and typically $M_Q/M_{ev} = = 0.70$ (Cogan, 1970; Rodgers, 1970; Iben and Tuggle, 1972).

3. Beat Cepheids

The majority of population I Cepheids have a singly periodic light curve. However, there exists a subset of Cepheids of short period whose light curve is modulated. These Cepheids are indistinguishable from population I Cepheids with regard to their gravity and effective temperature (Rodgers and Gingold, 1973). The modulation may be interpreted as a superposition of two modes of radial pulsation which correspond to fundamental and first overtone (Stobie, 1970). Eight beat Cepheids have been analysed for their periodicities and they have similar properties. The period, P_0, lies in the range 2 to 4 days and the period ratio, P_1/P_0, lies in the narrow range 0.703 to 0.711 (Stobie and Hawarden, 1972). A knowledge of P_0 and P_1/P_0 is sufficient to calculate the mass of a beat Cepheid, M_b.

Equation (1) may be rewritten

$$Q_0 = \alpha R^{1/4} M^{-1/4}, \qquad (2)$$

where $Q_0 = P_0 R^{-3/2} M^{1/2}$. Elimination of R from Equations (1) and (2) gives

$$M_b = \alpha^6 P_0 Q_0^{-7}. \qquad (3)$$

The value of Q_0 may be derived from the ratio P_1/P_0 and is almost model independent (Fitch, 1970). The observed ratio of $P_1/P_0 = 0.707 \pm 0.004$ leads to $Q_0 = 0.042 \pm 0.001$ from Equation (3) of Fitch (1970). With $\alpha = 0.022$, $Q_0 = 0.042$ and $P_0 = 3$ days the mass of a typical beat Cepheid is $M_b = 1.5$. This is a factor 3 lower than the expected evolutionary mass.

Similar results have been obtained (Petersen, 1973) by considering the position of the beat Cepheids in the $P_0 - P_1/P_0$ diagram. Rodgers (1970) and Schmidt (1972) also pointed out this discrepancy although they expressed it in a different form.

4. Bump Cepheids

Another method of determining the mass is to consider Cepheids with an observed bump in their velocity curve. Similar bumps have been found in the non-linear pulsation models (Christy, 1970). If the bump is defined as the *second* (not secondary) feature in the velocity curve, then the phase φ of the bump satisfies

$$P_0 \varphi = \beta R, \qquad (4)$$

where β is a constant ($\beta = 0.25$, Fricke *et al.*, 1972).

Eliminating R from Equations (1) and (4) we derive the bump Cepheid mass

$$M_\varphi = \alpha^{4/3} \beta^{-7/3} P_0 \varphi^{7/3}. \qquad (5)$$

With $\alpha = 0.022$, $\beta = 0.25$ and observed values of P_0 and φ, we find $M_\varphi/M_{ev} = 0.6$ (Fricke et al., 1972).

5. Transition Period

The results of non-linear pulsation models of population I Cepheids (Stobie, 1969) have indicated that the transition period between fundamental and first overtone pulsation occurs at $P_{tr} \approx 7$ days. This conflicts with the observed transition period of about 3 days, which is identified with the region of the beat Cepheids and with the transition between the low amplitude sinusoidal light curves and the asymmetric high amplitude light curves (Gaposchkin and Gaposchkin, 1966). It is not clear how serious this discrepancy is in view of the suggested finite width of the transition region (van Albada and Baker, 1973) and the relationship between the linear and non-linear transition regions (Cox et al., 1972). Further work, particularly with the non-linear pulsation models, is required to clarify the theoretical transition region and to understand what parameters affect it.

6. Discussion of Discrepancies

The discrepancy between M_Q and M_{ev} (Section 2) is the one most frequently discussed in the literature. A number of reasons have been advanced as to the cause of this discrepancy, e.g. mass loss, Hyades distance modulus, $T_e - (B-V)$ transformation and errors in pulsation models (see for example, Cogan, 1970; Rodgers, 1970; Fricke et al., 1972; Iben and Tuggle, 1972; Schmidt, 1972). We hope to show by consideration of the other discrepancies (Sections 3, 4 and 5) that errors in the pulsation models is the most likely explanation.

The discrepancies (M_Q, M_{ev}) and (M_b, M_{ev}) may be derived from the results of either linear or non-linear pulsation theory. However, the discrepancies (M_φ, M_{ev}) and (P_{tr}) depend entirely on the results of non-linear pulsation theory. The calculation of M_Q is dependent on knowing the absolute magnitude and unreddened color of a Cepheid and the relations transforming these quantities into luminosity and effective temperature. To this extent, the results in Section 2 could possibly be explained by errors in the distance modulus, reddening and transformation relations. However, discrepancies (M_b, M_{ev}), (M_φ, M_{ev}) and (P_{tr}) are essentially independent of these errors, i.e. altering L and T_e will not remove the discrepancy. Furthermore the three masses M_Q, M_φ and M_{ev} cannot all be made to agree by any change in L or T_e (Fricke et al., 1972).

Thus the discrepancies considered together indicate that something is wrong with present pulsation models both linear and non-linear. The discrepancy (M_b, M_{ev}) shows this most clearly as the principal uncertainty in Equation (3) is α. As this mass determination is most sensitive to the constants in the theoretical relations, it is not surprising that the largest discrepancy should occur here if the pulsation models are in error.

The results also show that mass loss by itself cannot explain all of the discrepancies because the masses derived from the pulsation model results (M_Q, M_b and M_φ) are themselves inconsistent. Hence although mass loss cannot be excluded, it is premature to invoke it as an explanation of any of the discrepancies until consistency is achieved.

7. Errors in Pulsation Models

By errors in pulsation models we do not mean that pulsation theory is incorrect. Indeed pulsation theory can explain so many properties of Cepheid variables that it is considered basically correct. Rather the error may lie in the input physics (e.g. opacity table) which will affect the model structure of the Cepheid.

A preliminary investigation of arbitrary changes in the opacity table (Stobie, unpublished) has revealed that it is possible to find a *single* opacity change which can account for all three mass discrepancies. The opacity change required is uncomfortably large being up to a factor 4 greater than the Cox and Stewart (1965) opacities. However, relative to the opacities of Carson *et al.* (1968) this factor would not appear so large as their opacities at some densities and temperatures are up to a factor 3 greater than those of Cox and Stewart (1965). It would be most interesting to recalculate the properties of pulsation models if and when a complete opacity table is available using the results of Carson *et al.* (1968).

References

Albada, T. S. van and Baker, N.: 1973, *Astrophys. J.* **185**, 477.
Carson, T. R., Meyer, D. F., and Stibbs, D. W. N.: 1968, *Monthly Notices Roy. Astron. Soc.* **140**, 483.
Christy, R. F.: 1966, *Ann. Rev. Astron. Astrophys.* **4**, 353.
Christy, R. F.: 1970, *J. Roy. Astron. Soc. Can.* **64**, 8.
Cogan, B. C.: 1970, *Astrophys. J.* **162**, 139.
Cox, A. N. and Stewart, J. N.: 1965, *Astrophys. J. Suppl.* **11**, 22.
Cox, J. P., Castor, J. I., and King, D. S.: 1972, *Astrophys. J.* **172**, 423.
Cox, J. P., King, D. S., and Stellingwerf, R. F.: 1972, *Astrophys. J.* **171**, 93.
Fitch, W. S.: 1970, *Astrophys. J.* **161**, 669.
Fricke, K., Stobie, R. S., and Strittmatter, P. A.: 1972, *Astrophys. J.* **171**, 593.
Gaposchkin, C. P. and Gaposchkin, S.: 1966, *Smithsonian Contrib. Astrophys.* **9**, 1.
Iben, I. and Tuggle, R. S.: 1972, *Astrophys. J.* **173**, 135.
Petersen, J. O.: 1973, *Astron. Astrophys.* **27**, 89.
Rodgers, A. W.: 1970, *Monthly Notices Roy. Astron. Soc.* **151**, 133.
Rodgers, A. W. and Gingold, R. A.: 1973, *Monthly Notices Roy. Astron. Soc.* **161**, 23.
Sandage, A. and Tammann, G. A.: 1969, *Astrophys. J.* **157**, 683.
Schmidt, E. G.: 1972, *Astrophys. J.* **176**, 165.
Stobie, R. S.: 1969, *Monthly Notices Roy. Astron. Soc.* **144**, 511.
Stobie, R. S.: 1970, *Observatory* **90**, 20.
Stobie, R. S. and Hawarden, T.: 1972, *Monthly Notices Roy. Astron. Soc.* **157**, 157.

DISCUSSION

Cox: I was absolutely amazed recently to see a paper by Böhm-Vitense where she tried to get the masses and helium content of some stars in globular clusters and she used Christy's blue edges and Iben's blue edges. The difference only is that Christy has a formula for the opacity and Iben has got a detailed table. By using these two sets of blue edges, she got the helium content in these globular clusters to vary a *lot*. I always thought that Christy had made a pretty good fit. My question is: have you obtained your $\delta \log \kappa$ from Christy's fit or from a very detailed table such as Iben has produced?

Stobie: It was relative to a coarse table which you produced, coarse in the sense that it is a coarse grid.

Cox: My point is – you are right. There seems to be sensitivity and I was amazed at it.

Stobie: We did originally consider just absolute constant changes, say just multiplying κ by a factor 2, and that makes virtually no change to the periods that you get for a given star of given mass and radius. You have got to make a change in the gradient before you got some substantial effects. The opacity change required to give agreement is up to a factor 5 different from the original opacity table and this is getting a bit drastic.

Rodgers: In changing the gradient of the opacity, have you done any non-linear calculations to see whether it changed the transition periods in the right direction?

Stobie: All these changes I considered, made absolutely no difference at all to the transition periods and that is one reason why I have been reluctant to publish anything, because I could not get anything to change that transition period and it always was coming out near seven days. The discrepancy for that must be some other cause which I do not understand.

Rodgers: Could I make one comment; saying that Dr Cox throughout this discussion has treated mass and helium content as two variables which you could play with and compensate. I would like to make a point, as a simple minded spectroscopist. To change helium abundances, in what Professor Eggen would call medium period young disc population variables, is no small matter to people who have made lots and lots of detailed analyses of helium abundances in B stars. I do not feel that the Kiel group, for instance, would view with similar cheerfulness, the idea that you can throw helium abundance around in the kind of way Dr Cox might like.

Buscombe: Isn't it different in the interior than it is on the skin?

Cox: I am not talking about very much. I am talking about the difference between 0.35 and 0.45. There is a recent Russian paper where he took Stobie's blue edges, and he matched and he got 45–50% helium in these stars. The same thing could be done with normal evolutionary masses which I have talked about this afternoon.

Stobie: I would place little reliance on the blue edges because there are problems with the boundary conditions. The boundary conditions were obviously so sensitive you have got to be sure that you do it consistently. The difference was just so much you could not say what the helium abundance was with any certainty.

Consider the case where there is an opacity change which increases with temperature – this may seem pretty arbitrary, but it would perhaps not be inconsistent with the opacity changes which Stibbs, Carson and Hollingsworth have been considering. Admittedly they have done no opacities computation in the actual temperature and density region here but they definitely found factors of 2 or even 3, at about 10^5 and 10^6 deg K – higher than Cox and Stewart opacities. If you believe that the opacities at lower temperatures are contributed mostly by hydrogen and helium, which we think we know well, you would expect then that there must be some gradient somewhere along here if you believe their results. Do you have any comment on that?

Cox: I guess the answer to that is that I don't believe every syllable. The opacities are very uncertain, there is no doubt about that.

THE TEMPERATURE SCALE FOR CLASSICAL CEPHEIDS

SIDNEY B. PARSONS

Dept. of Astronomy, The University of Texas at Austin, Tex., U.S.A.

Abstract. In the study of Cepheid pulsation and evolution, it is very important to be able to obtain effective temperatures from some easily observable spectral characteristic. A review is given of the several discrepant relations between T_{eff} and intrinsic $B-V$ index given for classical Cepheids and/or yellow supergiants by Rodgers (1970) and Parsons (1971) (the same as Kraft, 1961; Böhm-Vitense, 1972; Schmidt, 1972, and van Paradijs, 1973). All temperatures agree well at $(B-V)_0 = 0.4$ but diverge toward redder colors; at $(B-V)_0 = 0.9$ the values range from 4900° (Böhm-Vitense, for class Ia-Ib) to 5770° (Schmidt, for class I stars). In each case the analysis leading to the assignment of T_{eff} to a given color index is not completely satisfactory, so it is difficult to make value judgements. Comparison of relations between T_{eff} and spectral class shows similar discrepancies.

The author's 1971 analysis of $UVBGRI$ photometry of 101 supergiants and Cepheids was based on a single-valued relation between line-blocking in each wavelength interval and T_{eff}. The slope of the T_{eff} vs $(B-V)_0$ relation was determined à la Oke (1961) by maximizing agreement between the photometric and spectroscopic radius variations in two Cepheids, while the absolute scale was fixed from the F5 Ib star α Per. The adopted value of 6425° (with $E(B-V) \simeq 0.06$) is supported by recent unpublished calculations of Hβ, Hγ, and Hδ profiles which fit the observed profiles in α Per quite well (except within 0.3 Å of line center) for $T_{\text{eff}} = 6350 \pm 100$. For stars much cooler than this, though, the Balmer line wings are not strong enough to serve as accurate criteria.

Analysis of continuous energy distributions is hampered by line-blocking corrections and by interstellar reddening. From synthetic spectra computed from model atmospheres in collaboration with R. A. Bell, a good 'reddening-free' parameter was found for the Stebbins and Kron system:

$$Q_1 = (V-G) - 0.86(G-I).$$

The relation between this parameter and intrinsic Violet-Green index is nearly independent of gravity, abundance and microturbulence. In fact, it is found empirically that the Cepheids follow a single curve in the $V-G$ vs Q_1 diagram during their pulsation, with no evidence for looping except in the case of l Car. The shift in $V-G$, for a given Cepheid, to the intrinsic relation from the model atmospheres yields the absolute color excess, independent of the cluster method. The resulting $E(B-V)$ values confirm Parsons' 1971 results and average about 0.07 mag. smaller than other values in the literature.

The determination of reddening does not solve the T_{eff} problem, but if there is a unique temperature scale for Cepheids and if a given star is 0.07 mag. redder than previously thought, then that star is about 150° cooler. This goes part way toward removing the discrepancy between evolutionary and pulsational masses, even keeping the T_{eff} vs $(B-V)_0$ relation fixed. At present there are no strong arguments against continuing to use the Oke-Kraft-Parsons relation

$$\log T_{\text{eff}} = 3.886 - 0.175(B-V)_0,$$

which is not significantly different from Rodgers' relation and is intermediate between the extremes. However, one must recognize that not only the equation but also the inputs are subject to some uncertainty, and that the use of $\langle B \rangle - \langle V \rangle$ instead of $\langle B-V \rangle$ does not give a meaningful value of $\langle \log T_{\text{eff}} \rangle$.

References

Böhm-Vitense, E.: 1972, *Astron. Astrophys.* **17**, 335.
Kraft, R. P.: 1961, *Astrophys. J.* **134**, 616.
Oke, J. B.: 1961, *Astrophys. J.* **134**, 214.
Paradijs, J. van: 1973, *Astron. Astrophys.* **23**, 369.
Parsons, S. B.: 1971, *Monthly Notices Roy. Astron. Soc.* **152**, 121.
Rodgers, A. W.: 1970, *Monthly Notices Roy. Astron. Soc.* **151**, 133.
Schmidt, E. G.: 1972, *Astrophys. J.* **174**, 605.

DISCUSSION

McNamara: As you're aware, the various determinations of colour excess wander all over the place in a systematic fashion, yet a student by the name of Kent Phelps who's determining colour excesses from field stars in the vicinity of Cepheids, very carefully going round doing photometry, has determined the colour excesses now for about 7 or 8 stars. His colour excesses were compared to various observers and you'll be glad to hear that they agree with yours better than anybody else. But still they are systematically different from yours again by about two hundredths of a magnitude in $(b-y)$. The colour excesses are generally, I think he would conclude, overestimated; and yours also slightly overestimated.

Rodgers: I'd like to ask you what do you think of Schmidt who determined a temperature scale from fitting deblanketed $(R-I)$ to your models and put the temperature back up again. He makes the point in the paper, which I could find difficulty in accepting, that the increased accuracy in determining the blanketing of R and I makes the method superior to one dependent on $(B-V)$. The actual colour difference in $(R-I)$ is so small for stars of 6000° that observational errors become significant, it seems to me.

Parsons: I haven't seen this unfortunately but I'm sure $(R-I)$ should be a good criterion. That part of the spectrum was certainly included in the six colour fitting.

Schatzman: Are your models based on stationary atmospheres with constant gravity and temperature? Does it include a possibility of velocity gradient, passage of shock fronts, something like that in the atmosphere?

Parsons: No, it does not. They're completely static.

Iben: One drawback of changing the temperature scale is that if you fit in a mass-luminosity diagram, you leave a gap in the H-R diagram between the blue edges and the Cepheids, which would be as embarassing as a mass discrepancy.

Cox: But then you could vary the helium content. (laughter)

Parsons: Even by a change of 7 hundredths of a magnitude in $E(B-V)$, one does not probably change, move the Hyades. When Sandage and Tammann derived their P-L-C relationship, they used a ratio of total to selective absorption of 3.0 which is too low for these cooler stars.

uvby COLOURS OF CEPHEID VARIABLES

R. A. BELL

Dept. of Physics and Astronomy, University of Maryland, College Park, Md., U.S.A.

Abstract. Synthetic stellar spectra have been computed for Cepheids and supergiant F and G stars, using flux constant model atmospheres. These spectra have been used to compute theoretical *uvby* and *UBV* colours which may be compared with the observed colours. The supergiant α Per is used to establish the zero points of the theoretical colours. The colour comparison can be used to determine the temperatures, gravities and metal abundances of the stars as well as an estimate of the interstellar reddening. The reddening determinations are carried out by arguing that the temperature and gravity values obtained from $U-B$ and $B-V$ must match those obtained from c_1 and $b-y$.

A more detailed account of this work will be given elsewhere.

DISCUSSION

Tayler: Are not those gravity variations that you've got there really unbelievably large? A factor of 4 in gravity means a factor of 2 in radius.

Bell: Well, there's acceleration terms; it is an effective gravity, not a static one. The radius change would be so small that you would get practically no change in gravity from that effect alone.

Rodgers: Yes, you are in fact measuring a pressure here, and this is the pressure through the photosphere. That's what the effective gravity is about; that's where the factor of 2 is understandable.

NON-PULSATING STARS AND THE POPULATION I AND II INSTABILITY STRIPS

ARTHUR N. COX and JAMES E. TABOR

University of California, Los Alamos Scientific Laboratory, Los Alamos, N.M., U.S.A.

and

DAVID S. KING

University of New Mexico, Albuquerque, N.M., U.S.A.

Abstract. With specially computed detailed tables of equations of state and opacities, the instability strips for δ Scuti stars and Cepheids of population I and RR Lyrae and W Virginis stars of population II have been compared using the linear pulsation theory. Uncertainties in the observed strip locations and sometimes the mode of the observed pulsations do not allow high accuracy in fixing helium contents or the variables masses. Nevertheless, if masses close to those given by evolutionary theory are used, the helium content in population II objects is likely less than $Y=0.25$. A helium content of close to zero would put the theoretical blue edge of the instability strip to the red of the observed red edge, and have all the hotter stars which are in the strip as non-pulsating. For population I, Y can be more than 0.3, (more than 0.4 if half evolutionary masses are used), but if a given star has Y less than about 0.2 (full mass) or 0.25 (half mass), it can appear in the dwarf and classical Cepheid strips as non-pulsating.

DISCUSSION

Breger: Dr Cox presented a diagram in which some stars were pulsating and some were not. This is another diagram like that for all these stars which have been tested for pulsation. Near the main sequence, there are δ Scuti variables and non-variables. Indeed two thirds of the stars in the instability strip do not show any pulsation in excess of one hundredth of a magnitude amplitude. If the amplitude of any of these at the time of observation were larger than a hundredth, I think they would have been detected. I think there's quite a bit of evidence for non-variable stars or stars with very little variation, unfortunately you cannot say which of these points are actually completely stable. In other words, the fact that you don't see this progression or wandering of the blue edge across this diagram or see a lot of variables in the middle and very few at the left, may purely be a reflection of the fact that you cannot distinguish between real non-variables and variables with very small amplitude. We have looked for pulsation in six population II stars – old disc population stars – with very high proper motion. None of the 6 stars showed any pulsation whatsoever. This is not a very big sample, and we are continuing our investigation to see whether we do in fact find more in our AI Velorum type stars, but so far it is very disheartening and there may indeed be quite a few non-variables in the instability strip amongst these population II, low mass, stars.

Schatzman: Dr Cox's results raise two problems. In the case of the stars near the main sequence, A. Baglin has shown that it was possible that gravitational separation of helium could take place, in which case the stars are not pulsating. The stars in which, due to mixing, helium separation has not taken place, are pulsating. This would explain the presence of the 2 types of stars in the same area. For the case of Cepheids this seems to be much more difficult because they have had a long evolutionary sequence and the separation should have taken place much before the giant phase. Off hand, I would say it is more difficult to explain.

Cox: I, of course, did not discuss the question of inhomogeneous stars. Let me point out that my paper surely does not easily explain variables and non-variables in the same cluster. You see, for a given cluster in which one would hope to have the same composition for all the stars, there should be only one blue edge corresponding to the one helium content. If there is helium separation from the helium ionization or driving region, then non-pulsation can occur. Rotation can prevent the separation making pulsators, but the observations seem to indicate rotation prevents, not enhances, pulsation. It may be that gravitational separation of helium increases the helium in the driving region making pulsation. Then rotation would cause mixing, a net helium depletion and no pulsation. Cepheids have not much rotation and pulsate nicely.

Bell: In connection with the diagram which Dr Breger showed, I'd like to ask you if you can calculate models which show a light amplitude of say 1 or 2 hundredths of a magnitude?

Cox: No, we have not done that, I said the work here was all linear theory so all I can do is get a blue edge. I just know that it is pulsationally unstable at infinitesimal amplitude. We should study these with a non-linear theory and we will, but non-linear theory is so complex, it takes a long time to do it.

Zahn: What is the e-folding time of the amplitude instability in Cepheids compared to the time they spend in the Cepheid strip?

Cox: Anywhere from 10 to 10^6 periods the shorter growth times being for population II Cepheids and the longer growth times for δ Scuti stars. In any case, these times are always short compared to the time spent in the instability strip.

Bessell: Eggen and I observed a star last year which was variable over three nights of 3 hundredths of a magnitude; since then several others have observed it and found it not to vary. I have also observed several stars Danziger and Dickens suggested were variable and found no variations in them this year.

Cogan: What is the basis for stating that the position of the red edge of the instability strip is independent of chemical composition? You said that the radiative gradient was independent of chemical composition. Is this out of ignorance or do you have some reason for doing this?

Cox: I am told by some stellar structure people that they don't expect the onset of convection to be all that composition dependent, but unfortunately as we all admitted this morning, we don't really know exactly what it is about convection that causes the red edge.

Cogan: It is the onset of convection that causes the return to stability.

Cox: Yes. Alan Sandage suggested that if the helium moves the blue edge, it could also move the red edge and have the strip width constant. If reddening corrections and the conversion from color to effective temperature can be assured, the red edge might show a change with helium also. In this case we don't have adequate theory yet for the return to pulsational stability when there is convection.

Tayler: The results that have just been mentioned about the δ Scuti type stars, with the million period e-folding time are similar to the very high mass stars that I'm going to say something about to-morrow. In that case, since the higher harmonics are stable, you have a tremendous amount of time for interaction between the fundamental mode and the higher harmonics and it may be that although you think from a linear theory that you're going to get instability, you may get very little if the energy can be transferred by non-linear effects into the higher harmonics on a time short compared to the e-folding time. This is what we found in the high mass stars.

Cox: There is a problem about which harmonic theoretical blue edge corresponds to the observed blue edge for δ Scuti stars. All I know is that there is a blue edge defined for the fundamental, first harmonic, and second harmonic. I then inquire about what Q value the observers say corresponds to the observed blue edge. The Q value quoted is appropriate to the fundamental mode. Therefore I match the fundamental blue edge and get a reasonable helium content.

A further point is that if the helium fraction is lower than about 0.2 by mass, the big bang production value, then there is no pulsation for any star. These helium contents of 0.22 for population II stars and between 0.28 and 0.38 (or higher for less than evolutionary mass Cepheids) for population I stars seem reasonable. My paper is not able to get accurate helium contents due to the lack of precise observed blue edges.

Stobie: Surely this is a case where you could look at an external galaxy like the SMC where you can observe population I Cepheids and the question then is could the observations be of sufficient accuracy, could you detect any constant stars which are inside the strip?

Rodgers: You've got differential reddening that would foul the system.

Cox: The only way I have proposed to explain non-variable stars in the instability strip is to reduce their helium content which is causing the pulsation.

Iben: The only way you could explain away variables is to adopt a He abundance that's less than population II variables.

Rodgers: You had to go to a higher Y value for population I didn't you?

Cox: Yes, you see, to get a Cepheid to pulsate in the instability strip, I have to have a Y of about 0.3–0.35, or maybe if the mass is half the evolutionary mass, it's up to 0.5. I need considerable helium enrichment to make population I stars pulsate.

PULSATION MODES AND PERIOD-LUMINOSITY-COLOR RELATIONS FOR THE SHORT-PERIOD δ SCUTI STARS

MICHEL BREGER

University of Texas at Austin and McDonald Observatory

Abstract. Due to recent contradictory results, the period-luminosity-color relation of δ Scuti stars has been examined by a maximum likelihood method. δ Scuti stars are found to definitely obey a period-luminosity-color relation. The derived coefficients are very similar to those derived for RR Lyrae variables. Care has to be taken to allow for the large mass differences between the population I variables, such as δ Scuti stars, and population II ultrashort period pulsators.

The value of the pulsation constant Q is derived for a large sample of δ Scuti stars by calibrating the narrowband *uvby* indices in T_{eff} and $\log g$ from a complete set of model atmospheres (ATLAS). Most variables pulsate in the fundamental mode or the first overtone. Analogous to RR Lyrae stars, the division into different pulsation modes may be determined by color. For the main-sequence pulsators, this is not in agreement with theoretical transition period results which predict that no δ Scuti stars should pulsate in the fundamental (radial) mode. Should non-radial modes be important for δ Scuti stars, then the non-radial pulsation modes need to have Q values quite similar to radial modes.

DISCUSSION

Irwin: How were the original M_v's determined for these δ Scuti stars?

Breger: For cluster members, and there are quite a few of those, they were derived from cluster membership. For a few stars they were derived by a new calibration of *uvy β* photometry, which Crawford has made and which we have made.

Irwin: You said later that you used trigonometric parallaxes somewhere along the line.

Breger: Yes. For the dwarf Cepheids, because a photometric calibration assumes that you have a normal mass for the luminosity (I don't think you have a normal mass), photometric calibrations break down.

Rodgers: What are the effects of differential metallicity on the calibration, are they large?

Breger: You mean on the observed period?

Rodgers: No, on the calibration of effective gravity as a function of c_1.

Breger: I have not investigated this; I think Dr Bell can say something on that. I've used normal solar abundance models. The effects are likely not to be large, though. An Am star may be an extreme abundance star, in fact an Am star seems to fit these calibrations.

THE ABUNDANCES AND GRAVITIES OF THE δ SCUTI, AI VELORUM AND RR LYRAE STARS

M. S. BESSELL

Mount Stromlo and Siding Spring Observatory, Australian National University, Canberra, Australia

Abstract. A comparative investigation of the temperatures, gravities and abundances of the AI Velorum stars and some selected δ Scuti stars and RR Lyrae stars has been carried out at Mount Stromlo during the past few years. Eleven AI Velorum stars, 6 δ Scuti and 6 RR Lyrae stars together with 6 established standard stars have been observed. The eleven AI Velorum stars comprise almost all those accessible from the Southern Hemisphere. The RR Lyrae stars were selected as examples of the weakest and strongest line stars of type a and c, and the 6 δ Scuti stars observed also represent the strongest and weakest line stars of the group.

Spectra covering the region from $\lambda 3800 - \lambda 4600$ have been taken with an RCA 33016 image tube at 10 Å mm^{-1} dispersion. From these spectra, Hγ and Hδ profiles and the Ca K line equivalent widths have been measured. In the accompanying figure is

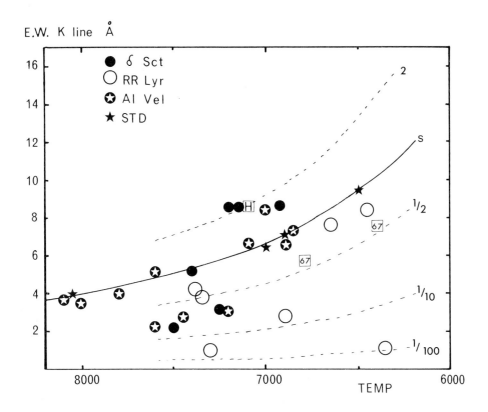

shown the equivalent width of the K line plotted against the effective temperature derived from the hydrogen lines. The continuous line represents the locus for solar composition stars and the dashed lines the loci for indicated multiples of the solar composition. The symbol H corresponds to a Hyades cluster member and 67, to two blue stragglers in M67. This figure shows clearly that there is no distinction in Ca abundance between the δ Scuti stars and the AI Velorum stars, and that the strongest line RR Lyrae stars have almost solar abundance. Photoelectric scans from $\lambda 3300$ to $\lambda 7500$ have also been obtained for all these stars. The effective gravities derived range from $\log g = 4.0$ to 3.2 and show that the δ Scuti stars and AI Velorum stars with period longer than 0.05 days are pulsating in the fundamental mode with values of $Q_0 \simeq 0.032$. There is again no separation between these two groups in the period $\sim \log g$ diagram. The RR Lyrae stars however do separate clearly, all having gravities between $\log g = 3.0$ and $\log g = 2.5$. In conclusion, on the basis of measured abundance, gravity and temperature there appears to be no clear distinction between the AI Velorum stars and the δ Scuti stars.

DISCUSSION

Breger: HD 24550 seems to be the star furthest down from your $\log P$ $\log g$ line. Derek Jones reobserved it and found a larger period which would put it on the line.

Bessell: Yes, and then the next year he observed it again and got it back below the line (laughter). There's just an interesting point that's got nothing to do with what I said before, but this star is 24550B and that's 24550, so there's a binary star in which one component is 2 times overabundant and the other component is normal. Derek Jones found this out also a couple of years ago, it is rather interesting.

McNamara: You do find the log of the gravities for the dwarf Cepheids varying between about 4 and about 3.5, is that about right?

Bessell: Yes, I found BS Aqr was the lowest with $\log g$ about 3.4.

McNamara: Do you have any comments about the possibility of a period luminosity relationship? Do you feel strongly about them?

Bessell: None of them are in clusters, we've only got a luminosity for SX Phe and if Eggen twists my arm a lot, I can take SX Phe up to 4.1 and get 1.2 solar masses on a 1/10th abundance star as a blue straggler; that's twisting everything to the extreme, but it is possible that that would fit in quite well.

THE DWARF CEPHEIDS

D. H. MCNAMARA and W. R. LANGFORD

Dept. of Physics and Astronomy, Brigham Young University, Provo, Utah, U.S.A.

Abstract. Intermediate-band photometry has been used to derive effective temperatures, metal abundances, and surface gravities of nine stars of the dwarf Cepheid group of variable stars. A strong correlation exists between the metal-line strengths and the period of pulsation in the sense that the shorter-period stars have weaker metal lines. The dwarf Cepheids are evidently found among both population I and population II stars.

The surface gravities and effective temperatures have been used in conjunction with pulsation theory to calculate masses and absolute magnitudes. The masses derived for the shorter-period stars are $\lesssim 1\ M_\odot$ in agreement with earlier estimates, but the longer-period stars are found to have masses in the range of 2–3 $M_\odot$. We also find that the dwarf Cepheids exhibit a period-luminosity relation.

The spectra of the dwarf Cepheids indicate that all of them have small rotational velocities, $v \sin i \leqslant 20$ km s^{-1}. We have also detected the Li line $\lambda 6707$ in the spectrum of VZ Cnc.

The observational evidence indicates that the dwarf Cepheids are post main-sequence stars that have not yet entered the red-giant stage, as suggested by Eggen, rather than stars in a post-red-giant evolutionary stage as was generally believed heretofore.

DISCUSSION

Breger: Where do you think the RR Lyrae stars start and the dwarf Cepheids stop?

McNamara: I have a feeling that the strong line RR Lyrae stars are simply an extension of this group. I'm not in a position to guarantee that, but I'm going to look for lithium in my strong line RR Lyrae stars, hopefully this fall, and I wouldn't be a bit surprised to find it.

Bessell: The problem with this metallicity criterion is that amongst the short period group there are some which do have young disc space motions and amongst the long period group there are some which have old disc space motions. So if you define them as old disc on the basis of the space motions then that picture doesn't hold; I agree with the abundances, but from the space motions the picture doesn't hold together too well.

McNamara: For this particular group that I've observed, at least abundancewise, the Δm_1 correlation seems very good. Of the stars that I had available I think there were only 3 with high velocities, and I haven't looked into that motion problem for the longer period groups.

Sinvhal: One of us recently worked on SZ Lyncis and came out with a mass of about 1.2 solar masses as a result.

McNamara: I see. I think that could agree very well with where I'd place it.

Stobie: What is your explanation for the high metal abundance group, why is there so few dwarf Cepheids compared to δ Scuti stars?

McNamara: I think it's the rotational velocity, again this may be just a classification problem as indicated this morning. If you plot the number of stars versus the rotational velocities, you find that there's a pretty flat distribution here for the δ Scuti variables. On the other hand, the stars of the type we've been talking about here with these large amplitudes all have very small $v \sin i$. The δ Scuti variables are distributed all along here, and some of the stars that have been classified as δ Scuti variables I think really belong to this group. And the other δ Scuti variables, a very small fraction that we observed that have small rotational velocities, I think, are stars simply that are rotating fast but have the $\sin i$ factor (projection factor) such that they're thrown in to the small $v \sin i$ values. I think the rotational velocity plays the key role here.

Breger: I think for δ Scuti stars this is indeed the case because the big amplitude δ Scuti stars are sharp lined giants usually.

Iben: I'm still a bit confused. I would think there'd be a certain amount of overlap between the low metal and metal-rich group in these diagrams.

McNamara: I would agree with you, because if you're taking the stars off a zero-age main sequence you expect some of these of lower metal content to be mixed in with some of higher metal content. I'm rather surprised at how sharp this relationship is.

Iben: But is there a split into two pieces or is this a continuous thing?

McNamara: I think it is continuous. That's the thing that shocks me. It's as if we worked our way down to various zero-age main sequences. Would you get some strong line ones, then, of extreme short period?

Bessell: Yes, HD 199757.

McNamara: I'm happy about this, not unhappy.

Rodgers: Do you see it as a problem though that the rate of growth of [Fe/H] is a function of time; $\delta \log t$ is not large. This just goes against every other piece of evidence we know about $\delta(\text{Fe/H})/\delta t$, for the old disc right up to the Hyades.

McNamara: I think it works out. I think this is what you expect.

MULTIPLE PERIODICITIES IN δ SCUTI STARS

R. R. SHOBBROOK

The University of Sydney, Sydney, Australia

and

R. S. STOBIE

Mount Stromlo and Siding Spring Observatory, Australian National University, Canberra, Australia

Abstract. Four δ Scuti stars (θ Tuc, ϱ Phe, 1 Mon and 21 Mon) have been observed intensively in an effort to understand the periodicities present in these stars. The stars selected were known to be of variable amplitude and relatively large light range. The data of each star was Fourier analysed. In no case did the period ratios have values consistent with the period ratios of low order radial modes. The most interesting result appears in the analysis of 1 Mon. The three principal periodicities present occur at $P_1 = 0.13612$, $P_2 = 0.13378$ and $P_3 = 0.13855$ days. To within the errors these periods form an equal frequency split, i.e. $1/P_2 - 1/P_1 = 1/P_1 - 1/P_3$.

DISCUSSION

Cox: I do not understand how you get these periods, I assume it's by some Fourier analysis?

Stobie: Yes, it is by Fourier analysis.

Cox: Therefore it assumes that these pulsations in all the modes are sinusoidal. Now if they are not sinusoidal, what does that do to them?

Stobie: For a start, you calculate your mean curve and see just how sinusoidal it comes out, and if it does not look sinusoidal, you'd expect a significant power to appear in one of the harmonics of this frequency at exactly half that period or one third that period, whatever it is. In fact, in this star we just looked at some results that Bob Shobbrook brought from Sydney and it turned out that for this star which is of large amplitude, there was evidence that there was a component I think at $0^{d}.06$ which has a very small amplitude. Most of these stars, the lower amplitude ones, have really sinusoidal curves when you look at them.

Schatzman: What is the meaning of a Fourier analysis for a saw toothed oscillation?

Stobie: Once you've got what you think is the main period, even if the thing is completely non-linear in shape, you can pick out a mean sine wave out of that.

Schatzman: I do not mean that you cannot do the calculation, I'm asking about the physical meaning of a Fourier analysis on a non-linear oscillation.

Stobie: But the assumption is that it's a combination of modes.

Schatzman: But this is not. This is a non-linear oscillation; not a superposition of linear modes.

Kemp: In physical systems, one can certainly have non-linear oscillators which can have a rigorously periodic behaviour, such that the oscillation waveform can be represented by a discrete Fourier series. (An ordinary pendulum for example is a non-linear oscillator, for finite amplitudes, yet is still strictly periodic, actually multiply periodic, if dissipation is negligible).

Rodgers: Can I ask Dr Cox or Iben about the linear combination of various modes? Has anyone done anything like this and is there anything physically in the numbers game that is indulged in, in s distortions and ϕ distortions of dwarf Cepheid light curves, like AI Velorum? What does the non-linear pulsation calculation say about them?

Cox: I don't know. I looked at that recently and I was quite appalled, I didn't think it really was legal doing that and I sort of feel that way here but your amplitudes are lower and therefore more sinusoidal.

Stobie: This would not add anything that had any physical meaning, it was a purely mathematical device to reduce this non-sinusoidal wave by phase distortion.

Cox: That is correct, I don't know the answer really.

Schatzman: You have observed for 59 nights one of the stars, the question is whether with this analysis you can predict something concerning the phase one year ahead, and I think you can not.

Shobbrook: The process is merely to look at the frequency spectrum of the observations and pick out sine curves, but we also of course look at the harmonics. Three have very small first harmonics, half the main period, and they are the three highest peaks in the frequency spectrum.

Savedoff: Is it correct to say that out of 10^4 data points, by taking out 6 parameters you have extracted most of the information and the residuals have dropped from 16/100ths mag. to 2/100ths mag.?

Stobie: Yes.

Schatzman: Are these periods just the result of the combination of non-linear oscillations, the periods of which *have* some physical significance? You're going to have harmonics of these non-linear oscillations of large amplitude and these harmonic periods will have no physical significance. For example, the harmonics you obtain will not describe the number of nodes from the centre to the surface of the star.

Rodgers: But in beat Cepheids you can analyse the harmonic periods and obtain period ratios which do have physical significance.

Sinvhal: We observed a δ Scuti star in two runs separated by two years and found the periods of the major Fourier components to be unchanged over the 2 year interval.

Stobie: Yes, this is the sort of thing that Fitch has done with similar results.

Warner: The equal spacing in this frequency domain is exactly what you would expect from non-radial pulsation involving a rotational perturbation and I have done a quick calculation which shows the rotation period to be 1 day or 2 days generating 3 modes, either 0 ± 1 or 0 ± 2 and a distinguished friend here tells me that $v \sin i$ is about 25 km s^{-1} which would give a rotation period of 1 day.

Stobie: I would have thought that the distortion required to produce this effect would have been quite large and would require rapid rotation.

Shobbrook (added in press): There is some confusion in the foregoing discussion regarding *harmonics* and *overtones*. The Fourier components of a non-sinusoidal periodic wave are *harmonics*, or *integer* multiples of the fundamental frequency which merely define the (constant) shape and amplitude of the wave. *Overtones* are other possible modes of vibration of the star, having a non-integer relationship with the fundamental, and whose interaction with the fundamental will produce a beat phenomenon.

β CANIS MAJORIS STARS IN THE (β, $[c_1]$) PLANE

R. R. SHOBBROOK

The University of Sydney, Sydney, Australia

Abstract. Recent publications have discussed the existence of stable B stars in the β CMa instability strip of the (β, Q) plane (Lesh and Aizenman, 1973) and the ($\log g$, θ_e) and (M_{bol}, $\log T_e$) planes (Watson, 1972). The new lists of Strömgren $uvby\beta$ photometry published by Crawford *et al.* (1970, 1971a, 1971b) for most of the B stars in the *Bright Star Catalogue* suggested that one might determine quite precisely the proportion of stable B stars (that is, stars not known as β CMa variables) in a luminosity/colour plot in the form of the β index against $[c_1] = c_1 - 0.2 (b-y)$.

The rectangle in Figure 1 shows the region of the (β, $[c_1]$) plane enclosing 22 of the 25 known β CMa stars. The variable lying in the top, left hand corner of the Figure is V 986 Oph, that just above the rectangle is τ^1 Lup (which has an uncertain β value) and that well below the rectangle is 53 Ari, probably with an incorrect β value. The dots in the figure are single early B stars brighter than $V = 5.0$, plotted to

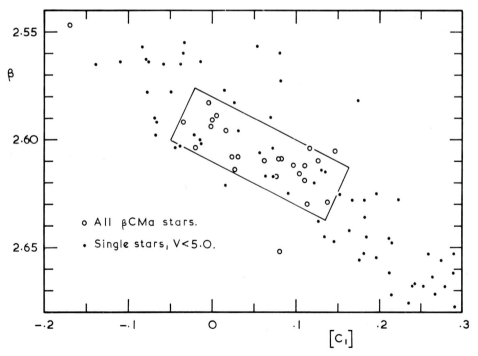

Fig. 1. A (β, $[c_1]$) plot of the known βCMa stars (open circles) with β and $[c_1]$ values corrected for effects of a companion when applicable, plus single B stars brighter than $V = 5.0$ (dots).

show the distribution of stars throughout the diagram; only 18 of the β CMa stars are brighter than $V = 5.0$ and 13 of the 25 are double stars.

In the three Crawford et al. lists, there is a total of 56 stars which lie in the rectangle ('instability strip') in Figure 1; these do not include stars classified as 'e' or 'p', since none of the β CMa stars is so classified. Both the β and $[c_1]$ values for double stars (including double β CMa stars) were corrected to the values for the primary by means of simple empirical relations. Where magnitude differences between components are not known, it was assumed that the secondary is 2–3 mag. fainter than the primary; the corrections are accurate to about 0.003 in β and 0.01 in $[c_1]$.

There are two items of information we can derive from this plot. Firstly, if we take into account that about 12 p.c. of early B stars are classified as 'e' or 'p', then a minimum of 37 ± 11 p.c. of the stars evolving through the instability strip (as defined by the rectangle) are β CMa variables. Secondly, the strip is not resolved by the observational data. This is inferred from the fact that the rms deviation of the β CMa stars in β about a line through the centre of the rectangle parallel to the long sides is ± 0.008, whereas the photometric errors in β are stated to be between 0.005 and 0.008. If we add to the latter values the errors in β determined for the primaries of double systems and the errors in $[c_1]$, we can see that the width of the rectangle is very largely due to the photometric errors. Similar conclusions are reached for the best available (β, Q), (M_v, Q) (Jones and Shobbrook, 1973), $(\log g, \theta_e)$ and $(M_{bol}, \log T_e)$ relations.

The most direct observational method of reducing the width of the instability strip significantly, so that there is a chance of resolving it in the 'luminosity' parameter, appears to be the reduction of observational errors in β. In principle, if errors in β can be reduced to 0.002 or 0.003, one might expect the width of the strip to decrease and the proportion of β CMa stars in it to increase. Moreover, if the 56 stars mentioned above are re-observed in β to this accuracy, we presumably obtain a short list of stars which may be undiscovered β CMa variables. However, although the currently known β CMa stars are highly concentrated to the rectangle in Figure 1, it will also be important to check for β CMa-type variability amongst the stars in the top left of the diagram (towards V986 Oph) and also in the bottom right amongst the $B2.5$ and $B3$ subgiants and giants.

References

Crawford, D. L., Barnes, J. V., and Golson, J. C.: 1970, *Astron. J.* **75**, 624.
Crawford, D. L., Barnes, J. V., and Golson, J. C.: 1971a, *Astron. J.* **76**, 621.
Crawford, D. L., Barnes, J. V., and Golson, J. C.: 1971b, *Astron. J.* **76**, 1058.
Jones, D. H. P. and Shobbrook, R. R.: 1973, *Monthly Not. Roy. Astron. Soc.* **166**, 649.
Lesh, J. R. and Aizenman, M. L.: 1973, *Astron. Astrophys.* **22**, 229.
Watson, R. D.: 1972, *Astrophys. J. Suppl.* **24**, 167.

DISCUSSION

Graham: Is there any chance that some of your stars have hydrogen line emission? It is what one would suspect.

Shobbrook: The spectra do not show emission.

Irwin: What is your purpose in observing β more thoroughly? Is it to reduce the thickness of the instability box?

Shobbrook: I think that careful observations on a few stars will eliminate up to 3/4 of the stars from this box.

Rodgers: Do the $(P\sqrt{\varrho})$ values from your observations differ significantly from those found by Watson?

Shobbrook: The calculated periods for β Cen, λ Sco, κ Sco and α Vir differed by 0.1 in log P from the observed values but there's nothing systematic, nor any difference for narrow or broad line stars.

Breger: Watson also calculated a box in M_v. Does it have the same width as yours?

Shobbrook: His box in M_v is about the same size as the one I found. The log g is a bit difficult because there's a lot of colour dependence.

Lesh: The stars which lie outside the β Cephei box in your diagram should not cause too much concern. V986 Oph is a very bright star, but it may not be a β Cephei star. As I recall, there is no radial velocity curve for it. As for 53 Ari, recent observations have shown it to be constant in light. But I am puzzled as to the location of τ Lup in your diagram. Crawford's published value of β is >2.7 for this star. Do you have a more recent value?

Shobbrook: Moreno in his Sco-Cen association study gives a value of 2.605.

A SEARCH FOR β CEPHEI PULSATION IN EXTREME COMPOSITION MODELS

ARTHUR N. COX and JAMES E. TABOR

University of California, Los Alamos Scientific Laboratory, Los Alamos, N.M., U.S.A.

Abstract. Following Davey (1973), we have considered the linear radial pulsational stability of stellar models in the region of the HR diagram populated by β Cephei stars. Instead of one composition, however, we have considered many compositions, some of them rather extreme. No consideration of composition changes for temperatures above a few-million degrees was given, and therefore the stellar structure and its pulsational stability may not be evaluated very accurately. Modulation of nuclear reactions was also neglected. In no case was radial pulsational instability found.

Reference

Davey, W. R.: 1973, *Astrophys. J.* **179**, 235.

DISCUSSION

Iben: It seems to me that the key to the solution of this thing is that the observations put the stars right in the overall contraction phase rather than at the end of the main sequence phase and during that overall contraction phase there's a huge convective shell. The star is contracting like mad and flinging out all sorts of flux that makes a real mess; there is a jagged composition profile. Hasn't anybody bothered to examine the pulsation of the complete star in that particular phase?

Aizenman: Davey carried out a linear non-adiabatic analysis in all phases of evolution of the star – the core burning stage, overall gravitational contraction phase, and thick shell phase. He found that the star was stable. I have carried out similar computations for stars of 10 $M_\odot$ and 15 $M_\odot$ during the same evolutionary stages and I obtain results similar to those of Davey.

Iben: But you didn't have time dependent convection, this thing will be jostling around all over the place. In fact it could be that fluctuation of convective motions act as the continual perturbation that drives the pulsation.

PART III

LARGE MASS STARS, STABILITY IN SUPERGIANTS, CRITICAL MASSES

NON-LINEAR INSTABILITY OF STARS WITH $M > 100\ M_\odot$

J. C. B. PAPALOIZOU* and R. J. TAYLER
Astronomy Centre, University of Sussex, England

Abstract. Massive main sequence stars are vibrationally unstable to small perturbations excited in the region of nuclear energy generation. This instability may, however, be limited at finite amplitude. Only the fundamental radial mode is unstable with an *e*-folding time of 10^5 to 10^6 pulsation periods; in contrast the overtones have damping times short compared with the growth time of the fundamental. Non-linear effects couple the linear modes. At very low amplitude the energy transfer time between modes is long compared with both growth and damping times and, as the overtones decay, a pure fundamental should result. However, as the amplitude increases, the transfer time becomes less than the growth and damping times. Energy can then be transferred from the fundamental to the overtones and damping can lead to a limitation of amplitude.

As integration for 10^6 periods is impossible, this behaviour cannot be verified by direct calculation. Several authors have integrated modified equations in which the excitation and damping have been speeded up considerably and they have found that non-linear effects limit the instability. They have not, however, found that overtones play any role. Papaloizou (1973a, b) has used the same technique to obtain an initial non-linear solution but he has then integrated the unmodified equations. He has found that, if the amplitude is high enough, overtones rapidly appear and, moreover, the most prominent overtone is one whose frequency is close to being an integral multiple of the fundamental frequency. It appears from his calculations that stabilization occurs at an even lower amplitude than previously suggested.

He has since studied non-radial modes in massive stars, although he has not yet performed detailed non-linear calculations. He has found that radial and non-radial modes in rotating massive stars are strongly coupled even for quite slow rotation speeds. This is particularly true for the overtones where the frequencies of radial and non-radial modes are very close and this coupling could be very important in the non-linear development of the vibrational instability.

Although the excitation mechanism in δ Scuti variables is very different from that in the massive stars, this is again a situation in which the *e*-folding time is 10^5 to 10^6 periods. Once again it is possible that there will be complicated behaviour due to coupling between the fundamental and the overtones, with non-radial oscillations playing a role in rotating stars.

---

* Present address: Department of Astrophysics, Oxford, England.

References

Papaloizou, J. C. B.: 1973a, *Monthly Notices Roy. Astron. Soc.* **162**, 143.
Papaloizou, J. C. B.: 1973b, *Monthly Notices Roy. Astron. Soc.* **162**, 169.

DISCUSSION

Cox: Does Papaloizou get larger or smaller limiting amplitudes for these massive stars than Talbot, Appenzeller or Ziebarth?

Tayler: The amplitude would be rather lower: the stabilization would occur at a lower amplitude than that predicted by Talbot, Appenzeller and Ziebarth. Papaloizou gets more rapid stabilization due to this transference of energy to the overtones, but he generally confirms the results of Appenzeller, Ziebarth and Talbot. I can't remember what his amplitude was, because I can't remember what the mode looks like. What I can remember is that when the ratio $\delta r/r$ at the centre is 0.04, he is already getting damping of the oscillation. Having got this thing into a finite amplitude you can't watch long enough either to see it grow or decay, so in order to decide whether it's decaying or growing, he has to use a generalisation of the energy integral that's used for the linear oscillations to find what the *e*-folding time or decay time is. Using that, he predicts that the thing is already decaying at one amplitude and growing at an amplitude a bit below that.

Iben: Didn't Appenzeller suggest that mass was shot off in these objects as a result of shockwaves?

Tayler: Papaloizou would say that as he gets a limiting amplitude which is rather lower than the other people, these effects might be rather less. He hasn't done a very detailed atmospheric calculation; no more than anybody else has. I think the estimates that Appenzeller and the others made of mass loss was done rather crudely. That's not a criticism, it's a comment on how difficult it is to try and put a proper atmosphere on to a pulsating star of any sort, let alone one of these rather awkward ones.

Aizenman: If the computations were to be carried out for times long enough for the overtones to die away, would the behaviour of the light curves then become similar to those of either Appenzeller or Ziebarth?

Tayler: He was worried about that very point. Any transients due to the overtones will not decay for a very long time, but the point that makes him relatively confident in his results, is that the only overtones he gets are always the overtones which are closest to resonance. If it was purely transience, he would expect it to be an arbitrary overtone. In other words, if you change the mass of the star then you find a different overtone is close to resonance, and as you crank up the mass of the star, you first get one bump on the ascending light curve, then 2 then 3 then 4 and then as you crank up the mass again, it starts going down again because the modes come into resonance in the opposite order. It would be rather surprising that you would have got this regularity if it was purely transient.

Van Woerden: Do you have any estimates for the amplitude of the light curve?

Tayler: I can't remember.

Aizenman: This is really directed to Art Cox. In your non-linear computations of Cepheids, do you find any behaviour similar to that described in this talk before the star begins to oscillate in a pure overtone?

Cox: No, we don't see this kind of thing.

Stobie: Is there any reason why the periods that you've calculated are from linear theory and not from non-linear calculations; is it because of the great length of time of the decay period?

Tayler: I think so.

Sargent: Might I ask what kind of stars in nature these objects are supposed to correspond to?

Tayler: The P Cygni's for instance, might be at the lower end of our model sequence, but as I say, it started off really as a fundamental theoretical problem: is there any reason why there shouldn't be more massive stars than are observed? That was the original idea. Well, now this would suggest that if there *aren't* more massive stars, it's either because the mass function peters out or because star formation will not allow such massive stars, and I know Larson has published papers in which he said you couldn't get a star this massive formed anyway. But, of the real stars, the P Cygni's are the ones that are nearest this region.

Cox: Is this paper disputing Larson's results?

Tayler: No, because this paper isn't considering the formation, only considering what would happen if you once got one.

Bessell: If the current evolutionary models of John Robertson are correct, then the stars in the LMC which have visual magnitudes around 10th magnitude are likely to be above 65 solar masses. These stars are about 2 mag. brighter than the 65 $M_\odot$ stars in Scorpius.

Tayler: Do they have any irregularities in their behaviour?
Bessell: They certainly do.
Tayler: They do – good (laughter).

ON THE VARIABILITY OF SUPERGIANTS OF TYPES B–G

A. MAEDER and F. RUFENER
Geneva Observatory, Switzerland

Abstract. The variability of the B–G supergiants is studied here with a set of 500 observations made for 76 stars. In addition to the discussion on the variability in colour indices (Maeder and Rufener, 1972), we also include the analysis of the apparent magnitudes. The observations, made randomly in time over several years, are selected from the new Catalogue (Rufener, 1974) of stars measured in the Geneva Observatory system. Some results on the luminosity dependence of the variations are presented in Table I.

TABLE I

Comparison of supergiants with normal MS stars

Luminosity class	N	Δm_v	ΔC	p
Ia	22	0^m063	0^m024	9
Iab–Ib	35	0^m045	0^m021	6
II	19	0^m027	0^m017	5
As a comparison: Normal MS stars		0^m023	0^m018	6

N is the number of stars observed in each luminosity class. Δm_v and ΔC are the amplitudes of the variations in V magnitude and colour indices. These amplitudes are defined as the range containing 90% of the individual measurements for a given star under the hypothesis of a normal distribution of the deviations to the mean. The quantity p is the mean number of individual measurements for each star. Table I clearly shows that the amplitude of the variations is increasing with increasing luminosity. From the study of the distribution of the amplitudes in each class, we may also infer that all supergiants of class Ia are variable in light. The stars of class II are essentially non variable for our threshold of sensitivity to variations. The stars of class Iab and Ib are intermediate.

The spectrum variations of the supergiants have been studied by several authors. Many spectral features, such as blue displaced circumstellar lines, reveal the existence of mass loss for these stars. However, indications of pulsation-type motions, besides the mass loss effects, have been given by Abt (1957) and especially by Chentsov and Snezhko (1971). From Figure 1, where luminosity is plotted vs period (or cycles) collected in the literature, we may draw further arguments in favour of the existence of some pulsation mechanism in supergiants. Figure 1 shows that the position of stars in the period-luminosity diagram is (a) spectral type dependent and (b) rather parallel

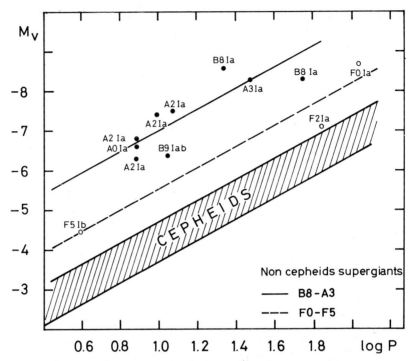

Fig. 1. Position of some supergiants in the period-luminosity diagram.

to the period-luminosity relation for Cepheids, as given by Sandage and Tammann (1968). These two facts may well be explained in the following way. If the variations are due to some oscillatory motion, one has the relation: $P \sim Q \bar{\varrho}^{-1/2}$ which gives, along with the mass-luminosity relation and the definition of T_{eff}: $\log P = \log Q - 3 \log T_{\text{eff}} - 0.24 M_{\text{bol}} + C$. Q characterizes both the stellar structure and the oscillatory motion. The above relation accounts reasonably well for the posision of the early supergiants relatively to the Cepheids in the P-L diagram and for the slope of this relation. The similarity of the Q-value for B-G supergiants and Cepheids is also well supported by the study of Figure 1. This suggests that the supergiants are subject to radial pulsation in the fundamental mode.

These facts (cycle-luminosity-amplitude relations) give some indications on the association of pulsation and mass loss in the supergiant star located in the HR diagram to the left of the instability strip of the Cepheids.

References

Abt, H. A.: 1957, *Astrophys. J.* **126**, 138.
Chentsov, E. L. and Snezhko, L. I.: 1971, in M. Hack (ed.), *Supergiant Stars*, 3rd Colloquium on Astrophysics held in Trieste, Osservatorio Astronomico, Trieste, Italy, p. 51.
Maeder, A. and Rufener, F.: 1972, *Astron. Astrophys.* **20**, 437.
Rufener, F.: 1974, *Astron. Astrophys. Suppl.* (in preparation).
Sandage, A. and Tammann, G. A.: 1968, *Astrophys. J.* **151**, 531.

DISCUSSION

Rodgers: Are there significantly non-periodic variations that are in excess of the errors that you have quoted?

Maeder: We have undertaken a set of long time base measurements of supergiants near the double cluster h and χ Persei, and it is measurements such as these that will answer your question.

Breger: The variability of HR 7308, an F supergiant, confirms Dr Maeder's comments. More than 100 observations show a fairly constant amplitude with a semi-period. The variations do not appear irregular, except for the slight difficulty with the period.

Underhill: You mentioned observing the variations of Be stars as well as the supergiants. Do you think the variations are of the same type?

Maeder: No, clearly no.

RECENT STUDIES OF THE BRIGHTEST STARS IN THE MAGELLANIC CLOUDS

PATRICK S. OSMER

Cerro Tololo Inter-American Observatory*, La Serena, Chile

Abstract. The objective prism surveys carried out by Sanduleak (1968, 1969) make available for the first time a relatively complete and homogeneous list of the brightest stars in both Magellanic Clouds for the range of spectral types O–G5. Photometric observations in the four color, $H\beta$ system of the 169 candidates in the SMC and the 45 brightest stars in the LMC as well as spectroscopic observations of the most luminous early-type stars, lead to the following picture of the stability and evolution of very massive stars:

(1) The reddening is the same for the most luminous stars in both Clouds and there is a sharply defined upper limit to the stellar luminosity, which is also the same in both Clouds. This limit corresponds to $M_{bol} = -11 \pm 1$ or $90\,\mathfrak{M}_\odot$, in good agreement with the classic study of Feast *et al.* (1960).

(2) The stars of highest total luminosity are the blue supergiants, and in both Clouds the high temperature limit corresponds to type B1.5 or $T_e \sim 25\,000$ K for $M_v \sim -8$. The SMC appears to be deficient in luminous F5–G5 supergiants.

(3) With few exceptions the most luminous stars are not known to show more than low amplitude variations in light. They do show $H\beta$ emission as expected, and four stars in the LMC have anomalously strong emission, indicative of unusual atmospheric activity. Bright Wolf-Rayet stars have long been known in the Clouds; in addition, several candidate Of stars are indicated by the photometry.

(4) The brightest B-type supergiants in the SMC appear to be deficient in N, Si, O, and probably Mg and C, although helium is normal, a result worth pursuing in the less luminous stars.

(5) Judging from the numbers of the brightest stars, the recent rate of star formation in both galaxies per unit mass is quite similar despite the fact that the SMC has a larger fraction of neutral hydrogen and the LMC a more conspicuous display of H II regions.

References

Feast, M. W., Thackeray, A. D., and Wesselink, A. J.: 1960, *Monthly Notices Roy. Astron. Soc.* **121**, 337.
Sanduleak, N.: 1968, *Astron. J.* **73**, 246.
Sanduleak, N.: 1969, *Cerro Tololo Inter-American Obs. Contr.* No. 89.

* Operated by the Association of Universities for Research in Astronomy, Inc., under contract with the National Science Foundation.

DISCUSSION

Bessell: In view of your comment that these stars are likely to be helium-burning stars, I would like to draw your attention to John Robertson's models of a B0Ia supergiant of 60 $M_\odot$, still in the hydrogen-burning phase. For older clusters in the Magellanic Clouds with turn off masses of 15 $M_\odot$, the existence of blue supergiants is probably due to the blue straggler process.

Osmer: I think you have to consider the time scale problem too, though. It must put some kind of restriction on how long they can burn hydrogen. I suppose this question will be answered by the number of main sequence stars we do find.

Hyland: Can you explain the observed slope of the upper limit of observed luminosities of Magellanic Cloud stars in the $M_{BOL} - \log T_e$ diagram?

Osmer: I think it's real, in that we surely know whether there are stars in this region of the diagram. However, the bolometric corrections are imprecise.

Irwin: Is there any information as to binaries or percentage of binaries in those stars near the upper limit of luminosity in your diagram. It would be nice to know the percentage of binaries and, of course, individual masses of some of these stars.

Osmer: There is very little evidence that I know of. I agree, it would be nice to know what fraction of the stars are binaries.

Irwin: I'd like to make just a comment. Many years ago, Shapley identified many of the brightest red stars in both clouds; he called these stars superluminous stars.

Przybylski: You said that stars in the Magellanic Clouds appear to have normal helium lines. There is one danger that one defines the spectral type of hot stars according to the strength of helium lines and in fact it may well be that there may be some stars which are actually hotter but helium weak. We cannot exclude this possibility.

Osmer: That was a good question. I just happened to be ready for it. I looked at that quite carefully in these B stars. I was concerned about whether this was all consistent or not. Si IV lines are quite strong in B0 and then drop right off quite quickly; Si III lines come up at B1, and then you get Si II coming up at B2. With the ratio of the silicon lines you can really pin down the temperature because things happen quickly here and then on the same scale, the other lines – let's say helium or carbon or nitrogen – most of them vary rather slowly. The final bit of information is of course the colour. I found that in fact you could get a consistent colour and spectral type from the silicon lines and then examine whether the helium and carbon and nitrogen and so on, were normal. Everything did work out consistently and the helium was identical to the stars in the Large Cloud and the one in Scorpius as far as I can tell. I agree about your point, but in these particular stars I think there is some reason to believe the helium is normal.

Parsons: Have you looked at the m_1 index, and if so, do you notice any systematic difference between the SMC and LMC?

Osmer: I have looked at the m_1 index, and so far, I have not notice any systematic differences but I've not pursued it too much. It's a difficult question for these very luminous stars simply because of a lack of good calibration. I think there's some useful data there; I haven't gone into it too far other than to note that it appears to be similar but I think you'd have to work at it fairly hard to get a meaningful result out of that.

VARIATIONS IN THE LINE SPECTRUM OF A-TYPE SUPERGIANTS

WILLIAM BUSCOMBE

Northwestern University, Evanston, Ill., U.S.A.

Abstract. Attention is drawn to a group of stars of luminosity class Ia, in which variation of radial velocity has long been known in semi-regular cycles of 5 to 30 days. The interval of approximate repetition appears to be positively correlated with the luminosity of the star.

Recent spectrophotometry of the absorption lines, especially by Rosendhal, Aydin, Groth, and Buscombe shows a considerable variation in the strength of Balmer lines (with emission components detected at Hα), as well as those of metallic ions. The microturbulence parameter derived from curves of growth for Fe II varies by a factor of two or more during some cycles.

DISCUSSION

Hutchings: From many spectra of 6 Cas taken during 1971–72 at Victoria, we find no indication of the radial velocity variations which Abt found in 1957.

Bell: Does the turbulence remain constant on your plates?

Hutchings: We have not obtained microturbulent velocities for all the spectra, but the few we have do not show much variation.

Buscombe: Perhaps I could just add that on the Radcliffe plates, as well as my own from Stromlo, it seemed that the strength of Hγ in absorption is well correlated with the microturbulence parameter and ranges over a factor of 2.

Underhill: Supergiants are very complex and it will be difficult to make model atmospheres. For instance, η CMa, B5Ia, shows evidence for a stationary circumstellar shell and a layer which is moving outward with a velocity of 100 km s^{-1}. How are you going to combine these in one model in hydrostatic equilibrium?

Buscombe: When I corresponded with Dr Thackeray, he urged me to take account of the fact that this Carina supergiant is the only one of the southern stars in my program which is involved in visible nebulosity and, indeed, the other three stars seem to be stable and constant in all respects.

PART IV

HALO AND OLD DISC POPULATIONS

A. RED VARIABLES AND EVOLUTION ON THE GIANT BRANCHES

RED VARIABLES OF THE OLD DISC AND HALO POPULATIONS

M. W. FEAST

Radcliffe Observatory, Pretoria, South Africa

Abstract. This paper discusses the following topics: (1) Mira Variables in Globular Clusters; (2) infrared observations of Globular Cluster Variables; (3) carbon Stars in Magellanic Cloud Clusters; (4) the Se Variables and the superlithium rich Variables.

Red variables represent an important but little understood phase of stellar evolution. Primarily for this reason, they have attracted an increasing amount of attention in recent years. This paper is restricted to a discussion of three or four aspects of this topic which have been of particular interest to a group of us working at the Radcliffe Observatory.

1. Mira Variables in Globular Clusters

The subject of Mira variables in globular clusters was discussed fairly extensively at the Toronto Colloquium last year. Let me briefly summarise the present position. The Miras in the general field (stars with Me spectra) show a marked dependence of kinematic property on period (e.g. Feast, 1963; Feast *et al.*, 1972). If we include with these variables, the Fe-Ke variables (the SRd variables of the General Catalogue of Variable Stars), we can form a kinematic sequence running from halo type objects with periods near 100 days down to disc objects of low velocity dispersion with periods of about 450 days (Feast, 1965; Preston, 1967). Examples of SRd Miras are found in some metal poor globular clusters; a result consistent with the halo kinematics of the SRd Miras in the general field. Seven globular clusters, all in the metal rich category, contain Miras with typical Me spectra. Until recently, there were only six periods known for stars in this group (three of these were in 47 Tuc). All these periods were near 200 days. This is consistent with the kinematics of field 200 day Miras which are intermediate between disc and extreme halo population types. Recent work (Lloyd Evans *et al.*, to be published) has shown that two other clusters have Miras of longer period. NGC 6553 has a Mira of period 270 days and NGC 5927 one of 312 day period. The latter star is most likely a cluster member from its position in the centre of the cluster and its radial velocity is consistent with membership. There is some evidence (Feast, 1973; Lloyd Evans and Menzies, 1973) that these clusters may be more metal rich than the 47 Tuc type clusters which contain the 200 day variables (spectral type near giant tip, $(V-I)$ colours of variables). These results therefore tend to reinforce the idea that there is a fairly precise relation of period with chemical composition and/or age for the Mira – SRd sequence. Most globular clusters containing Miras also contain smaller amplitude red variables which presumably show us how stars become pulsationally unstable as they approach the red giant tip and evolve with increasing amplitude and period, finally becoming Mira variables.

2. Infrared Photometry of Globular Cluster Variables

A more detailed study of the sequence of events amongst variables at the red giant tip is desirable. As one contribution to such a study, Glass and Feast (1973) have been carrying out broad band infrared photometry in the 1.2–3.5 μ region (J, H, K, L) of red giants and variables in 47 Tuc and ω Cen. The following is a summary of some of the results obtained.

The two colour $(J-H)$–$(H-K)$ plot yields some useful results. This type of plot is interesting because in it the population I luminosity classes I and III stars are well displaced from the black-body line. This is primarily due to a maximum in the stellar continuum in the H band (1.6 μ), caused by an opacity minimum in the stellar atmosphere at this wavelength. At longer wavelengths, H^- free-free opacity dominates and H^- bound-free at shorter wavelengths (Woolf et al., 1964, etc). It is found that the majority of the 47 Tuc and ω Cen stars studied fall near the position of the population I stars, well away from the black-body line. However, a striking feature of the globular cluster results is that the three 200 day Mira variables in 47 Tuc (V1, V2 and V3) lie much nearer the black-body line than the other stars. Fragmentary data on field Miras (Mendoza, 1967) indicate that their behaviour is similar to that of the 47 Tuc Miras. These results are presumably connected with the fact that, unlike other M stars, Miras show strong H_2O absorption bands in the infrared which depress the 1.6 μ continuum peak. At present, it is something of a mystery why Miras have strong H_2O absorption bands and other M stars of similar temperature do not (Johnson et al., 1968). Abundance differences have been suggested or it could result from unspecified differences in atmospheric structure. In any event, the 47 Tuc results seem to show that the effect comes in quite abruptly for the Miras. In the $(J-H)$–$(H-K)$ plot, the SR-small amplitude red variables in 47 Tuc lie with normal M stars.

Although M type stars have long been known in 47 Tuc, it was generally supposed that such stars would be confined to metal rich globular clusters. It was therefore quite a surprise when a few M type variables with periods of about 100 days and amplitudes in the range $0^m.4$ to $1^m.4$ were found in the metal poor globular cluster ω Cen (Dickens et al., 1972). The J, H, K, L photometry leads to some interesting results for these stars. It is found that the relation between the infrared colours and the spectral types as determined from TiO band strengths is the same as for field population I stars. Despite this, there are good reasons to suppose that these stars are, like other cluster members, metal poor, since they have large $(U-B)$ excesses ($\sim 0^m.4$) for their types and infrared colours. It has generally been assumed that, in giants of a given temperature, the TiO bands would be a sensitive indicator of metal abundance. On the contrary, our results show no dependence of TiO strength on metal abundance. A likely reason for this is that, in the spectral region where we study the TiO bands, TiO is itself a major source of opacity so that, to a first approximation, band strength is independent of metal abundance. Detailed calculations are needed to test this hypothesis.

The infrared colours indicate that the normal red giant branch in ω Cen ends at $T_{\text{eff}} \sim 3800°$. At this point, the stars have $\log L/L_\odot = 3.23$ and their positions agree well with the first giant branch tip (onset of the helium flash) calculated for $Z = 10^{-3}$ (Rood, 1972; Demarque and Mengel, 1973). This value of Z is consistent with that estimated for ω Cen in other ways; $(U-B)$ excess, abundance analysis of Fehrenbach's star (Dickens and Powell, 1973). The M type variable members lie at considerably cooler temperatures ($\sim 3100°$) but at about the same luminosity ($\log L/L_\odot = 3.16$). Since the helium flash occurs at a point on or close to the Hayashi line, we deduce that these variables are beyond the Hayashi line and presumably therefore in a stage of rapid evolution probably connected with mass loss.* It would obviously be of importance to decide whether these variables are on their first or second ascent of the giant branch. There is a strong temptation to believe that they are at the top of the giant branch for the second time. One may then maintain that these stars are showing the first signs of instability which leads to the ejection of a shell, the rapid evolution back across the HR diagram and the formation of a planetary nebula. Whilst this may indeed be the case, it appears to leave little place in the scheme of things for the two CH stars in ω Cen.** These carbon stars lie near the tip of the giant branch proper and an attractive suggestion would appear to be that they represent the tip of the second giant branch whilst the M variables are in a rapid excursion to the right of the Hayashi line from the tip of the first giant branch. If the M variables are indeed first giant branch stars, then they might well take care of the mass loss necessary to provide a spread of masses on the horizontal branch. A decision between the two possible alternatives for the M variables will evidently rest to a large extent on detailed calculations of stellar evolution at the helium flash.

3. Some Problems of the Carbon-rich Red Variables

To understand the evolutionary status of variable carbon stars (and carbon stars in general), we badly need to know their bolometric magnitude and temperatures. At the present time the temperature calibration seems to be particularly uncertain. Bessell and Youngbom (1972) have found that the temperatures of N type stars derived from continuum scans in the region 7000–11000 Å are lower than those derived from $(R-I)$ measures calibrated from M stars (2500 to 2900 K rather than 3300 to 3500 K). This is presumably due to the effects of strong CN absorption. However it is not certain that even Bessell and Youngbom's work refers to the true continuum. Evidently a great deal will need to be known about the infrared spectra before broad band colours can be properly interpreted.

An interesting problem in connection with carbon stars is the nature of some of the

---

* The results of Glass and Feast (1973) leave open the possibility that in 47 Tuc, the Mira variables may lie relatively close to the Hayashi track.
** A third carbon star has recently been reported in the region of ω Cen but this has been found to be a radial velocity non-member (Catchpole and Feast, 1973).

very red stars in globular clusters in the Magellanic Clouds which distinguish these clusters from globular clusters in our own Galaxy. Several of these very red stars in SMC globular clusters have now been examined spectroscopically (Feast, 1973; Feast and Lloyd Evans, 1973). The reddest star in Kron 3 $(B-V=2.57)$ is a carbon star. There is at least one carbon star and probably more in NGC 419 and the small amplitude variable V8 in NGC 121 is a carbon star. No such very red carbon stars (variable or non-variable) are known in galactic globular clusters (the CH stars in ω Cen being much bluer). However recent work suggests that red carbon stars may be rather frequent in intermediate age clusters in the Galaxy, (see the discussion of Catchpole and Feast (1973)). Of the small number of known intermediate age clusters, three (NGC 7789, 2660 and 2477), all with ages near 1.5×10^9 yr, are now know to contain carbon stars. This result suggests that the occurrence of these stars is rather sensitive to age and gives some attraction to the hypothesis that the SMC globular clusters may be somewhat younger than typical galactic globular clusters. Gascoigne (1966) and others have indeed suggested on other grounds that Kron 3 and NGC 419 might be intermediate age clusters but the situation remains rather confused.

In discussing small amplitude carbon variables such as NGC 121 V8, it should be borne in mind that whilst none are known in galactic globular clusters, such stars may be present in the galactic halo population. The majority of known carbon (CH) stars in the halo population have $(B-V)$ in the range of about $1^{m}\!.0$ to $1^{m}\!.5$ like the CH stars in ω Cen. However Table I lists two small range carbon stars with colours rather

TABLE I

Star	Period (days)	Approx. B amp.	$B-V$	U (km s^{-1})	V (km s^{-1})	W (km s^{-1})	ϱ	b
NGC 121, V8	112	0.4	1.7–2.15					
TT CVn	Irr	0.7	+1.85 (max)	−28	−353	−74	−135	+79.2
V Ari	77	1.0	+2.1 (max)	+107	+19	+345	−176	−45.7

similar to NGC 121 V8. Also listed are the components U, V, W of their space motions (Eggen, 1972) and their observed radial velocities and galactic latitudes. Both stars appear to be halo objects and presumably indicate that very red carbon stars are not entirely lacking in some component of this population.

4. The Se Variables and the Super Lithium Rich Variables

The Se variables are an intriguing and little understood group of stars. The limited data on their radial velocities suggest that they cover a wide range of ages since the majority show clearly the effects of differential galactic rotation (e.g. Feast 1963) whilst one Se variable is known with a radial velocity of 329 km s^{-1} w.r.t.l.s.r. (Catchpole and Feast, 1971) making it a halo object. The period of this latter star is 251 days (Andrews, 1973) which does not clearly distinguish it from Se variables of the low velocity group.

The Se variables may be of importance in helping solve the problem of the superlithium rich stars. It has been suggested (Cameron and Fowler, 1971) that the superlithium rich stars occur as a result of a brief phase of stellar evolution involving extensive convection down to a helium burning shell. In addition it has been suggested that the lithium formed in this way is a major source of ^{7}Li in the universe (e.g. Truran and Cameron, 1971).

Keenan (1967) and Boesgaard (1970) have found the Se variable T Sgr to be superlithium rich with a lithium abundance as great or greater than the T Tauri stars. Recently, a survey has been carried of lithium in southern S stars (Catchpole and Feast, to be published). Out of 169 S stars observed in this programme, only one is in the superlithium rich class. This is the Se variable RZ Sgr which is rather similar to T Sgr. Since our 169 S stars contain only 27 Se stars, the frequency of superlithium rich Se stars may be quite high, possibly higher than amongst carbon stars despite the fact that for many years the only known superlithium rich stars were the carbon stars WZ Cas, WX Cyg and T Ara. T Ara was found in a survey of 50 carbon stars (Feast, 1953). Warner and Dean (1970) observed another 158 carbon stars without finding another superlithium rich star. Recently about 70 more carbon stars have been observed (mostly by Catchpole) without finding any superlithium rich stars.

Superlithium rich stars are apparently particularly frequent in the intermediate CS and SC classes with two superlithium rich objects in the small group of only 18 known stars, most of which are SR variables, (Catchpole and Feast, 1971). The SC group, with 12 stars in it, is particularly interesting. These stars are all very similar to one another spectroscopically except that one has a very strong lithium line (E.W. ~ 3 Å). Infrared photometry (Glass, to be published) shows that the SC stars form a compact group in their J, H, K, L colours. The infrared colours of the superlithium rich SC star place it with the other SC stars and appear to rule out the possibility that the great enhancement of lithium in this star is a temperature effect.

The mechanism to produce superlithium rich stars discussed by Cameron, Fowler and others is expected to operate in relatively massive stars (~ 5 solar masses or more) and not to be very effective at lower masses. With only 7 stars known in the superlithium rich group, it is not possible to deduce much on this point from their kinematics except that there are no very large (halo type) radial velocities amongst the seven. However, our new superlithium rich Se variable RZ Sgr has the relatively high velocity of -38 km s^{-1} w.r.t.l.s.r. in a direction where the effects of differential galactic rotation are small and lies at a galactic latitude of $33°$ and so is probably well out of the plane. Velocity and position thus suggest either a moderately old low mass object (contrary to theoretical expectation) or possibly a run away star.

In this paper some aspects of current work have been discussed for each of the three main abundance classes amongst red variables, the M, C and S type stars. It will be apparent that in each group there are many unsolved problems requiring further observational and theoretical work.

Acknowledgements

I am grateful to R. M. Catchpole, T. Lloyd Evans and P. J. Andrews for discussions and unpublished data.

References

Andrews, P. J.: 1973, *Quart. J. Roy. Astron. Soc.* **14**, 451.
Bessell, M. S. and Youngbom, L.: 1972, *Proc. Astron. Soc. Australia* **2**, 154.
Boesgaard, A. M.: 1970, *Astrophys. J.* **161**, 1003.
Cameron, A. G. W. and Fowler, W. A.: 1971, *Astrophys. J.* **164**, 111.
Catchpole, R. M. and Feast, M. W.: 1971, *Monthly Notices Roy. Astron. Soc.* **154**, 197.
Catchpole, R. M. and Feast, M. W.: 1973, *Monthly Notices Roy. Astron. Soc.* **164**, 11P.
Demarque, P. and Mengel, J. G.: 1973, *Astron. Astrophys.* **22**, 121.
Dickens, R. J., Feast, M. W., and Lloyd Evans, T.: 1972, *Monthly Notices Roy. Astron. Soc.* **159**, 337.
Dickens, R. J. and Powell, A. L. T.: 1973, *Monthly Notices Roy. Astron. Soc.* **161**, 249.
Eggen, O. J.: 1972, *Astrophys. J.* **174**, 45.
Feast, M. W.: 1953, *Coll. Astrophys. Liège* **5**, 423 (Les Processus Nucléaires dans les Astres).
Feast, M. W.: 1963, *Monthly Notices Roy. Astron. Soc.* **125**, 367.
Feast, M. W.: 1965, *Observatory* **85**, 16.
Feast, M. W.: 1973, in J. D. Fernie (ed.), 'Variable Stars in Globular Clusters and Related Systems,
Feast, M. W. and Lloyd Evans, T.: 1973, *Monthly Notices Roy. Astron. Soc.* **164**, 15P.
Feast, M. W., Woolley, R. v.d. R., and Yilmaz, N.: 1972, *Monthly Notices Roy. Astron. Soc.* **158**, 23.
  IAU Colloq. **21**, 131.
Gascoigne, S. C. B.: 1966, *Monthly Notices Roy. Astron. Soc.* **134**, 59.
Glass, I. S. and Feast, M. W.: 1973, *Monthly Notices Roy. Astron. Soc.* **163**, 245 and **164**, 423.
Johnson, H. L., Coleman, I., Mitchell, R. I., and Steinmetz, D. L.: 1968, *Commun. Lunar Planetary Lab.* **7**, 83.
Keenan, P. C.: 1967, *Astron. J.* **72**, 808.
Lloyd Evans, T., and Menzies, J. W.: 1973, in J. D. Fernie (ed.), 'Variables Stars in Globular Clusters and Related Systems', *IAU Colloq.* **21**, 151.
Mendoza, E. E.: 1967, *Bol. Obs. Tonantzintla Tacubaya* **4**, 114.
Preston, G. W.: 1967, *Publ. Astron. Soc. Pacific* **79**, 125.
Rood, R. T.: 1972, *Astrophys. J.* **177**, 681.
Truran, J. W. and Cameron, A. G. W.: 1971, *Astrophys. Space Sci.* **14**, 214.
Warner, B. and Dean, C. A.: 1970, *Publ. Astron. Soc. Pacific* **82**, 906.
Woolf, N. J., Schwarzschild, M., and Rose, W. K.: 1964, *Astrophys. J.* **140**, 833.

DISCUSSION

Mavridis: I have two questions. First, did you determine the absolute magnitudes of the carbon stars in the Magellanic Cloud clusters? Second, can you tell us whether these stars are M type or F type stars?
  Feast: The absolute magnitudes I can tell you. As to your second question, I would not like to say.
  Mavridis: Otherwise tell us whether they are early C or late C.
  Feast: I can tell you that, in these two clusters at any rate, if you determine the intensity of the carbon bands, these are about 3 on the Morgan and Keenan's scale (which runs from 0 to 10). Their absolute magnitudes are about -3, which is somewhat brighter than the ones in the intermediate age galactic clusters in the galaxy, which are I think of the order of -2.
  Buscombe: The opacity, according to Hollis Johnson of Bloomington, Indiana, is more contributed by water vapour and by carbon monoxide than it is by TiO.
  Feast: In the 5000 Å region?
  Buscombe: No, not there.
  Feast: One wants to know what the opacity is in the region where you see the TiO in the spectrum.
  Buscombe: He has models in which he imitates the observed spectra with computation based on opacity and he has been quite successful with it.

Vardya: The H_2O opacity operates around one micron. Thereafter the CO will also operate.

Feast: I do not know whether Dr Vardya would like to say whether it seems reasonable that TiO could provide a reasonable amount of opacity.

Vardya: It may. The thing is H^- should go down and then the TiO can come up. That is very difficult to say.

Bell: I wanted to make one or two comments about the CH stars. Dickens and I have analysed the spectra of one of them in ω Centauri and we get a carbon 12 to carbon 13 ratio which is 10 instead of 90 in the solar case. We also get a higher nitrogen abundance than for the other elements which you expect to have an abundance of about 1/20 of solar. The comment I wanted to make to start with, is that your diagram is a little bit misleading in that the CH stars are reddened by the carbon features in their spectra. They are in fact occurring at something like 0.1 or 0.2 mag. redder in $(B-V)$ than their temperature would indicate.

Feast: We got infrared observations of at least one of these and they do not lie anything like as far out as the M stars in ω Cen. They are somewhere in the region of the Giant Tip.

Bell: Another question I would like to ask you is: in the spectrum of your star in Kron 3, which is your 16th mag. star, can you see any evidence for the existence of carbon 13? Can you see the C13-C13 head at about 4754 Å?

Feast: I cannot tell you at the moment. I think one would have to work on the spectrum fairly carefully, which I have not done. It is a very interesting point.

Bell: Another point is related to the star which you are giving a $(B-V)$ of 2.4. Such a $(B-V)$ could possibly be produced by excess carbon blanketing as well. Because if one can produce something like one or two tenths of magnitude in the ω Cen CH stars which have a relatively low C/O ratio, then if you take a high C/O ratio you could get very much greater carbon blanketing by CN and CH and consequent reddening of the star.

Feast: I am sorry, I certainly did not mean to imply that I thought the $(B-V)$'s of carbon stars told you what the temperatures were. It is just of course that in the galactic globular clusters you can use the $(B-V)$ to see whether you find stars like this. Could I ask Dr Bell a question? In the analysis that you did with Bob Dickens, can you tell the difference between a carbon over-abundance and an oxygen under-abundance?

Bell: It depends upon the temperature of the star because the strength of the C_2 features is very strongly dependent upon the temperature. It seemed that, in ω Centauri for example, you could not quite produce strong enough C_2 features if you took away all the oxygen. But of course the problem is that you do not really know what the oxygen abundance is.

One other question I would like to ask. Do you see any possibility of seeing carbon 13 features in your variable M stars. That could help you to decide whether or not they are on their first or second ascent of the giant branch.

Feast: Not in the spectra... If one went into the infrared it might be possible; they are not very bright stars, of course. It is a very interesting point.

Rodgers: Two points: one with respect to your comment about TiO supplying its own opacity. Ms Hain, now Mrs Shinkawa worked on S Carinae, and in working on field stars took $(R-I)$ observations and near infrared scans of continuum and band strengths in what she called low and high velocity giants. Professor Eggen made exactly the same observations and the results that come out of both of these are that the TiO band strengths are independent of what kinematics you assign to the star. On your picture it says either TiO is the dominant source of opacity or, for the high velocity stars, electrons for H^- opacity are deficient. This is likely to be true for supergiants but for the giants, the H^- opacity, is going down, and again you have this problem of if the TiO is down, what is the O doing.

The final point I want to make is that even among the early M type stars where TiO is patently not supplying its own opacity there is still no differentiation. The second point – you made the point that in the Miras in ω Cen you had in $\log T_e$ a red extension of the giant branch but no extention in luminosity. I would like to say that Dr Hyland and I are working on two Miras in NGC 6397, in which we think the metal abundance is down by at least 20 (this is the calcium abundance in the horizontal branch). These two stars lie two to three magnitudes below the giant branch in V, but in M_{bol} are way above the tip of the giant branch of that cluster.

Feast: These are Miras?

Rodgers: Yes.

Feast: Quite large amplitudes?

Rodgers: Yes, in V. The light curves are not well determined. I give a minimum of about 4/10ths.

Hyland: These are two stars which are right at the tip of the giant branch in NGC 6397 which on the $[V, (B-V)]$ diagram fall well below, as you said before. They were found to be variable by the Herstmonceux group and they did not give any period but you can get a rough sort of period between 100 and 200 days, and the large amplitude in V. In fact, when we first looked at one, it was 16th magnitude in V, in K it is third magnitude. So it is a pretty bright infrared source. Sixty days later, in V it was about 11th mag. Its K mag. had only changed by about 2/10ths mag. So this one is a really large amplitude in the optical but in the infrared it hardly varies at all, or has a very small amplitude. The other one is not as well determined as to what is going on.

The point about it, is that the two spectra of these stars with the TiO bands both show a spectrum around about M5. One of them is redder than the other and they have different calcium line strengths (4226 Å), which would indicate a difference in temperature. The one which is right up at the tip of the giant branch is much brighter than any in ω Cen or 47 Tuc; in fact one would expect it on its colour to have a spectral type of M7, M8... but it is still only M5 according to the TiO strength.

Przybylski: In addition to the comment made by Dr Rodgers I would like to stress one point concerning the TiO bands in late type stars in ω Centauri. Normally we would expect normal strength of TiO bands if oxygen is normal in the star. It is always so that if one component of a molecule has a normal abundance in the atmosphere then the strengths of the absorption bands does not change. That is what we observe, for instance, with CH bands. We all agree probably that in all kinds of stars, with a few exceptions only, hydrogen is normal and for this reason, the strength of CH bands is independent of the fact whether it is a star of population I or population II; exactly in the same way, for instance, Mg H is also normal in population II stars. If oxygen is normal we would also expect that TiO bands would be normal. The possible conclusion may be that in ω Centauri, oxygen is normal. Does anyone know whether there is any observation and evidence to the contrary? Incidentally, what is the strength of CN bands in ω Centauri?

Feast: I do not think that I can give you a qualitative figure. I would be very interested to hear from the experts on stellar atmospheres whether they would agree with that comment.

Rodgers: Peimbert reported on the composition of planetary nebulae at the IAU Symposium No. 58. In the planetary in M15 where Ca/H is down by thirty, oxygen is only down by a factor at 10. It still seems to us very hard to understand the strong TiO in the NGC 6397 Miras because we would have thought that with these extremely low gravity stars, Rayleigh scattering must play a fairly major role. They are extremely cold and Rayleigh scattering, I would think, must be competing – we have not done the sums yet – very very strongly with intrinsic band absorption by TiO. Of course, if you are right, we are underestimating the TiO abundance because the TiO band edges are in fact overlapping. It seems difficult to believe that these stars have the same surface composition as the rest of the stars in 6397.

Feast: I am fairly confident that the M stars in ω Cen are metal-deficient objects. Of course, you can always fix up the strength of the TiO by varying the elements in an arbitrary way, but the fact that you have got to get back to the relation you find for normal abundance stars makes it rather suspicious to just juggle the abundances.

Vardya: I was wondering why some of the M stars become very much like Mira variables, others do not. Are there any clues from the observation side which might tell us what makes a star become Miras where others do not.

Feast: I did not have it on those slides but we have got some infrared observations of V42 in ω Cen which is one of the largest amplitude red variables. It has got a period of about 150 days and it may very well, if one uses the Mira terminology, be a metal deficient Mira. In the observations we have got in a $[(J-H), (H-K)]$ diagram, it falls with the 47 Tuc Miras as though the opacity in the H region is being affected in this star in ω Cen, by water vapour. Of course, one has got to realise that these M stars in ω Cen are pretty rare. A lot of stars were gone through to find that.

DYNAMICAL MODELS OF ASYMPTOTIC-GIANT-BRANCH STARS

P. R. WOOD

Mount Stromlo and Siding Spring Observatory, Australian National University, Canberra, Australia

Abstract. The dynamical behaviour of the envelopes of four 0.9 $\mathfrak{M}_\odot$ asymptotic-giant-branch stars of composition $(X, Z) = (0.68, 0.02)$ and luminosity $\log L/L_\odot = 3.41, 3.60, 3.85$ and 4.14 has been studied. At $\log L/L_\odot = 3.41$, the envelope pulsates in the first-overtone mode and exhibits properties similar to those of the Mira variables. The more luminous models were all found to pulsate in the fundamental mode while simultaneously undergoing violent envelope relaxation oscillations. A significant amount of mass loss occurred from the model with luminosity $\log L/L_\odot = 3.60$. It is suggested that a redgiant star undergoing envelope relaxation oscillations would be recognised as a symbiotic star. In the two most luminous models, a distinct outward-moving shell containing almost all the hydrogen-rich material in the star is formed. A suggestion is made that a further increase in luminosity would completely eject this shell to form a planetary nebula.

DISCUSSION

Iben: I have a number of questions. The first one: Are these models complete? That is, do you construct an asymptotic branch star including the core or do you just pick the surface temperature and go down and stop at some distance from the center?

Wood: I got a complete evolutionary model, got the luminosity-core mass relation (there's a very well defined luminosity/core mass relation) and if you integrate inward, you find that at about a third of a solar radius, there is a big discontinuity in pressure. So there's a very well defined core, so all you really need from the stellar interior calculation is the luminosity/core mass relation and then you can apply the boundary condition where the pressure goes up very sharply and I chose my core masses to agree with the luminosity/core mass relation for asymptotic giant branch stars.

Iben: So evolutionary calculations *do* put the stars to the red of the first giant branch by a considerable amount.

Wood: Rood's Z was 0.01, mine was 0.02, I think that's the reason for the difference.

Iben: Number 2, what's the driving mechanism, H_2?

Wood: No, if you look at the driving mechanism, the driving occurs right at the top of the hydrogen ionization zone so it's linked in with hydrogen mainly and possibly convection.

Iben: Can you comment on stars of intermediate mass in this phase?

Wood: I've done other calculations at 1.2 $M_\odot$ and 1.5 $M_\odot$ and the effect of this is, you're not changing the luminosity/core mass relation, that's pretty independent of anything for these stars. What you're doing is putting all that mass into the envelope and the effect of that is to move the top of the hydrogen ionization zone closer to the surface in mass, and that means that the driving which occurs at the top of the ionisation zone is closer to the surface. That tends to favour first overtone pulsation or overtone pulsation for higher modes, so the effect is that if you increase the total mass you raise the luminosity at which the transition to fundamental pulsation occurs. And, as you saw, fundamental pulsation always seems to be associated with some violent behaviour, like mass loss.

Iben: Why is it that convection doesn't damp out the pulsation? Is there some point where convection, although it occurs, carries so little flux that it has no influence on the driving mechanism?

Wood: All the models I've done have convective envelopes.

Cox: But what fraction of the flux is being carried by convection?

Wood: Just about all of it.

Iben: Then how can you have a driving mechanism?

Wood: It all happens at the very surface of the ionisation zone and I think it may be something to do with matter passing in and out of this convection zone which starts at the hydrogen ionisation zone. It's a very sharp discontinuity at the beginning of convection.

Kemp: How do you include mass loss in the hydrodynamic model? Doesn't that produce a mathematical problem since the mass inside the envelope is not conserved?

Wood: I haven't. When things get too far away, you've exceeded the escape velocity. I just stop my calculations when things blow up. But there's no artificial mass loss put in, it just comes out of the dynamics.

Kemp: Is the part of the envelope that exceeded the escape velocity conserved in the calculation?

Wood: Yes, because you're considering the envelope as a whole and the matter that's escaped is still in the envelope. It's passed escape velocity but you've still got it in the dynamic calculation.

Schatzman: For such an extended envelope you are likely to have convective velocities which are highly supersonic, which means that the turbulence being supersonic, you have always conversion of the kinetic energy into heat and this complicates the picture a little bit for the convective zone. Have you taken that into account?

Wood: I had a look at the convective velocities and they didn't exceed the velocity of sound, I think because the specific heat is so high in hydrogen and helium ionisation zones you don't need a very large velocity in order to carry the required flux. If you go to hotter stars it does get supersonic.

Schatzman: That is very close to the surface but you have a zone which is convective over one half of the star or more than that.

Wood: Convection carries less of the flux at these deep levels.

B. SHORT PERIOD VARIABLES, HORIZONTAL AND ASYMPTOTIC BRANCH EVOLUTION AND PLANETARY NUCLEI

VARIABLE STARS AND EVOLUTION IN GLOBULAR CLUSTERS

P. DEMARQUE

Yale University Observatory, New Haven, Conn., U.S.A.

[Paper not submitted by the author]

DISCUSSION

Cox: May I ask about these period changes? There has been this problem (that maybe you have overcome, maybe you have not) about the observed period changes which are too rapid. What you are predicting is rapid negative changes, but not very rapid positive changes.

Demarque: That is right.

Cox: Have you compared that with the observations?

Demarque: It is very difficult to get anything consistent out of the people who make the observations.

Cox: My impression was that they had rapid changes but they were both positive and negative.

Iben: And of about the same amplitude. In ω Cen, they tend to be in one direction predominantly but they are all about a factor of 10 larger than evolutionary changes during the major core He burning phase.

Cox: But he has got the factor of 10 now, Icko.

Iben: No. If you look at these things, 10^6 yr were spent in that little spike out of 10^8 yr. That means only one out of 100 stars in the instability strip should be doing that, but they are all doing it.

Demarque: I think it would have to be demonstrated that such jagged composition profiles do exist and we would like to be able to establish a time scale of any changes in this profile.

Iben: It seems to me that a number of discrepancies that you have caused to come about could disappear if perhaps semi-convection was not quite as important as your calculations seem to indicate. That is, the lifetime argument suggests only the order of 10% helium, for example, whereas pulsation plus observations say something like 20%, and without semi-convection it says 30%. I would argue that this suggests that the truth lies somewhere in between leaving out semi-convection altogether and treating it to the extent to which you do. The next point is that the BL Her stars are certainly there and I think are more abundant than W Vir (or Pop. II Cepheids) by quite a bit and that the tracks that you argue now replace the superhorizontal branch stars do not seem to get over there far enough.

Demarque: The particular one you saw did not but of course there is a whole spectrum all the way to the star that will go up the asymptotic branch. The interesting thing about it is that, only those that will not become asymptotic branch stars will stay there long enough to become BL Her stars because the others move up the asymptotic branch too fast.

Tayler: To return to your semi-convection zone, is it not true that in the region which you are interested in, the mixing length is probably comparable with the radius? The mixing length is normally very small compared to the radius, but here the mixing length would be getting rather large and therefore in order to really understand what is going on you probably need a very much better idea of what convection is doing than in a simple mixing length theory. Maybe you get considerable irregularities in your semi-convection due to this sort of thing.

Demarque: The words 'semi-convection' are misleading. We do not treat any type of convection. We just say there is some mixing.

Tayler: What I mean is that if you are in a situation where you have got the scale height being of the same order as the radius you would expect that you might get irregularities in whatever goes on in the semi-convection which might lead to something which would complicate your story, maybe in the way you want it complicated.

Zahn: About semi-convection again, you refer to the ocean but in the ocean you have a situation with warm water on the top. It is colder underneath and the "salt" content is in the opposite direction. So you have a stable temperature gradient and an unstable salt gradient whereas in a star it is just the contrary, you have an unstable temperature gradient and a stabilizing μ gradient. This situation has been studied in the lab where you have heat below and have an unstable temperature gradient but you have a stabilizing salt gradient. What you get in the lab is a finite amplitude overstability which mixes the whole tank. It is a finite amplitude overstability which means there is a threshold for that and nobody knows what that threshold would be in a star. There is one difference, in the lab the whole situation is dynamically stable, there is a stream constant, whereas you start with a neutral equilibrium and maybe that case is special. So one might expect a mixing of the whole semi-convective region leading to a homogeneous region.

Demarque: That could have essentially the same effect as the composition instability mentioned earlier. I should add that this possibly may have a relation to the β Cephei phenomenon because there is reason to believe that the same effect occurs but at much lower central hydrogen abundance, at the time of the hydrogen exhaustion. It could be very interesting to study this line.

THE CHARACTERISTICS OF THE FIELD RR LYRAE STARS IN THE MAGELLANIC CLOUDS

J. A. GRAHAM

Cerro Tololo Inter-American Observatory*, La Serena, Chile

Abstract. The Magellanic Clouds are well known as being very suitable for observing the various stages of stellar evolution. During the last few years, I have been studying the RR Lyrae variable stars in each of the two Clouds. Some first results were reported at *IAU Colloquium* No. 21 in 1972 (Graham, 1973). Here, I would like to update these results on the basis of more recent data and to comment on some of the characteristics of the field RR Lyrae stars in each system. Periods and light curves are now available for 63 RR Lyrae stars in a $1° \times 1.3°$ field centered on the cluster NGC 1783 in the Large Magellanic Cloud (LMC) and for 62 stars in a $1° \times 1.3°$ field centered on the cluster NGC 121 in the Small Magellanic Cloud (SMC). Both ab and c type variables are represented and, viewed individually, the Cloud RR Lyraes are identical in characteristics to those known in our Galaxy. Studied as groups, however, there are small but significant differences between the RR Lyrae stars in each system. The following four specific features seem to be emerging from the study.

(1) The mean apparent magnitudes seem to cluster very closely about a single value. In the LMC, the time averaged mean visual magnitude $\langle V \rangle$ is $19^{m}_{.}20$ and in the SMC, $19^{m}_{.}65$. Using the best available distances for the Clouds (see Graham, 1973), these correspond to $M_{\langle V \rangle} = +0^{m}_{.}5$ and $+0^{m}_{.}45$ respectively, with a probable uncertainty of $\pm 0^{m}_{.}2$ in each value. These absolute magnitudes are somewhat brighter than those predicted by the simple application of the often used Christy (1966) formula to the values of the transition period in each case.

(2) ab type variables with large amplitudes and periods less than $0^{d}_{.}45$ are very rare or absent in both Clouds. Not one is found among the 125 stars studied. Such variables would need to have maxima fainter than blue magnitude 20.0 to have missed being detected in this study.

(3) There is an obvious difference in the period distribution and in the relative numbers of ab and c types. This is best summarized by noting that the transition period is $0^{d}_{.}48$ for the LMC sample and $0^{d}_{.}52$ for the SMC sample. c type variables seem to be approximately twice as common among the SMC sample as compared with that in the LMC. Following the trend seen in Galactic globular clusters, this seems to indicate an average metal deficiency of 2 or 3 in the SMC old population near NGC 121 as compared with that in the LMC sample. However, it is important to note that the RR Lyrae absolute luminosities appear to be about the same in both cases.

* Operated by the Association of Universities for Research in Astronomy, Inc., under contract with the National Science Foundation.

(4) There is considerably more dispersion in the LMC period-amplitude diagram than in the corresponding one for the SMC variables. In the LMC, there seems to be no correlation between the scatter of points representing individual stars and their apparent magnitudes. The scatter may be due to an age dispersion, a composition dispersion, or a mixture of both in the LMC field, but whatever the cause, the mean absolute luminosity of an RR Lyrae star does not appear to be affected by it.

References

Christy, R. C.: 1966, *Astrophys. J.* **144**, 108.
Graham, J. A.: 1973, in J. D. Fernie (ed.), 'Variable stars in Globular Clusters and Related Systems', *IAU Colloq.* **21**, D. Reidel, Dordrecht, Netherlands, p. 120.

DISCUSSION

Cox: Can you tell me what is the value for the transition period in the galaxy? I gather it's less than 0.45 days.

Graham: Yes. It depends on what part of the galaxy you look at and what sort of sample you take, but you do get many stars going right down to 0.4 days.

Cox: But not less than 0.4 days.

Graham: No.

Bessell: Derek Jones looked at these stars in our Galaxy shorter than 0.4 of a day and claimed that they probably had absolute magnitudes of about +1.5. Do your observations in the Clouds suggest a similar situation?

Graham: It's on the limits of my observations, but these things have large amplitudes, I'm surprised that I didn't discover them when they hit maximum. When you take ten plate pairs your completeness is pretty high. For the ab type stars it was over 90% in both cases.

Iben: How again did you obtain the distance moduli?

Graham: Various ways. Firstly I used Gascoigne's colour-magnitude diagrams for globular clusters; secondly I used the Sandage-Tammann PLC data.

Iben: Did those two give identical results?

Graham: More or less. There was a dispersion of 0.1–0.2 mag.

Iben: But there is still then the possibility that the modulus and the magnitudes are off by 0.2 or so?

Graham: At the most, yes. 18.7 is possibly the most stable modulus for the LMC.

Rodgers: With the mass-luminosity ratios appropriate to NGC 6397 and ω Cen HB stars, your RR Lyrae luminosities lead to a mass of 0.61 $M_\odot$. This seems a good number.

Graham: Yes.

ON SOME PECULIAR FEATURES IN THE SEQUENCES OF THE OLD OPEN STAR CLUSTERS

A. MAEDER

Geneva Observatory, Switzerland

Abstract. In spite of the rather good agreement between the theory of stellar evolution and the observations, there exist some difficulties when one compares closely the sequences of open star clusters and the theoretical isochrones. Several, if not all, of the old open star clusters seem to be concerned, especially those which are accurately measured, namely Praesepe, NGC 2360, 752, 3680 and M67. The problem concerns the gap occuring in the HR diagram at the end of the phase of hydrogen burning in the core; it corresponds to the phase of hydrogen exhaustion (or of overall contraction). The sequence of M67 has been studied by Racine (1971) and Torres-Peimbert (1971). The well apparent gap is located farther from the zero-age main sequence than indicated by the models and the hook towards a larger T_{eff} predicted during this phase is not observed. Differences in chemical composition may not be held responsible for these anomalies. From Torres-Peimbert's models, it may be assumed that neither solar type, nor super metal rich composition are able to reduce the discrepancies. As a further illustration, let us mention the case of NGC 752. In Table I, the main features related to the gap are examined: the disagreement, like in M67, essentially concern features 1 and 2. The observations are based on a recent

TABLE I

NGC 752: features of the sequence

Features	Observed	Observed on Bell's fig.	Iben (1967a) log age = 9.2	Hejlesen et al. log age = 9.2
1	1.75	1.60	0.75	0.85
2	0.30	0.35	0.30	0.30
3	−0.02	0.00	0.11	0.10
4	0.20	0.15	0.50	0.30

1 height (in magnitude) of the top of the gap above the main sequence.
2 vertical size (in magnitude) of the gap.
3 horizontal size (in $B-V$ index) of the gap. (Top minus base.)
4 height (in magnitude) of the subgiant peak above the top of the gap.

study of Grenon and Mermilliod (1973) and on Bell's data (1972). Bell has also mentioned the existence of discrepancies. As in M67, the gap is too far from the zero-age main sequence and does not present any sudden turning towards a larger T_{eff}.

It is well established (Iben, 1967b) that the presence of the gap depends in a one-to-one fashion with the occurence of convective mixing during the phase of hydrogen burning in the core. So, it is possible that the above anomalies are an indication of some modifications of convection in the stellar interiors relatively to what is generally assumed in computations of stellar models. In order to make some check on these points, we have tested the effects of overshooting of the convective core on the evolutionary tracks. In the present calculations, we have introduced this effect according to the discussion of Dilke and Gough (1972) about the solar neutrino problem, by limiting the ratio H_X/H_p to some critical value ζ. H_X and H_p are the scale heights of the hydrogen content X and of pressure p. In fact, other hydrodynamical processes may produce the overshooting of the convective core. Effects of this kind have been discussed for example by Spiegel (1972), Shaviv and Salpeter (1973). A common characteristic of these processes is that they are all producing a smoothing of the μ-gradient during evolution. Some results are shown in Figure 1. The smoothing of the μ-gradients tends to reduce the observational discrepancies. The gap occurs farther from the zero-age main sequence and does not present the large hook to the left. The case $\zeta=0.5$, which lies close to the observations, corresponds to a supplementary smoothing of the μ-gradient affecting a region of 6% of the total stellar mass above the con-

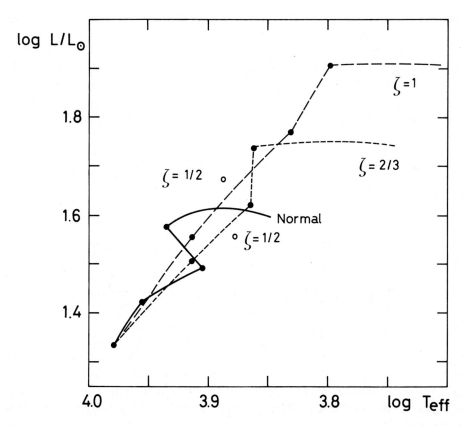

Fig. 1. Evolutionary tracks in the HR diagram for various cases of smoothing of the μ-gradients in a $2.25 M_\odot$ star. The track labelled 'normal' represents evolution computed under usual assumptions.

vective core. As more nuclear fuel is available during the phase of hydrogen burning in the core, such an effect leads to an increase of 40% in the duration of this phase.

References

Bell, R. A.: 1972, *Monthly Notices Roy. Astron. Soc.* **157**, 147.
Dilke, F. W. W. and Gough, D. O.: 1972, *Nature* **240**, 262.
Grenon, M. and Mermillod, J. C.: private communication.
Hejlesen, P. M., Jorgensen, H. E., Petersen, J. O., and Romcke, L.: 1972, in G. Cayrel de Strobel and A. M. Delplace (eds.), 'Stellar Ages', *IAU Colloq.* **17**, XVII-1.
Iben, I.: 1967a, *Astrophys. J.* **147**, 624.
Iben, I.: 1967b, *Ann. Rev. Astron. Astrophys.* **5**, 571.
Racine, R.: 1971, *Astrophys. J.* **168**, 393.
Shaviv, G. and Salpeter, E. E.: 1973, *Astrophys. J.* **184**, 191.
Spiegel, E. A.: 1972, *Ann. Rev. Astron. Astrophys.* **10**, 261.
Torres-Peimbert, S.: 1971, *Bol. Obs. Tonantzintla Tacubaya* **6**, 3.

DISCUSSION

Demarque: I was very interested in this star because Prather and I have been making very similar calculations where we've been worried about a discrepancy with observation for a long time. We have done it in a much more simple minded way, nothing as sophisticated as what you have done, and just looked into the effects of some slight amount of overshooting in a convective core. Of course you do get the same sort of effect, and in fact you find that if you do it this way that it is restricted to a region of the HR diagram which is more or less where M67 is. I would warn that the model of Dilke and Gough is controversial and I wouldn't want to tie my theory to their paper.

Maeder: Yes, I agree. My aim is mainly to show that a smoothing of the μ-gradient results in a modification of the evolutionary tracks in the sense required by the observations.

Iben: This is also going to have a tremendous influence on the position of core helium burning stars, in particular it will affect the mass-luminosity relationship. Have you carried the analysis in this direction?

Maeder: No, but the evolution of the red giant stars will probably be also affected by this criterion, because the internal history of the star is different when we come to the red giant region, but I did not look at this.

MODELS FOR NUCLEI OF PLANETARY NEBULAE AND FOR ULTRAVIOLET STARS

J. KATZ, R. MALONE, and E. E. SALPETER

Cornell University, Ithaca, N.Y., U.S.A.

Abstract. A series of stellar models were evolved, all with a total mass of 0.65 $M_\odot$, an initial carbon-oxygen core of mass 0.60 $M_\odot$, an intermediate helium mantle and an outer hydrogen-rich envelope with mass varying from case to case. Although the most hydrogen-rich cases resulted in red giants, cases with $\lesssim 0.01$ $M_\odot$ in the hydrogen envelope evolved at high surface temperature. The early stages of development of these models are similar to observed central stars of planetary nebulae. The later stages (when the nebula should be very thin optically) still have a high luminosity; the relevance to 'ultraviolet stars' will be discussed.

DISCUSSION

Demarque: Did you encounter any hydrogen shell flash?

Salpeter: Yes, some of these violent outward excursions were in fact hydrogen shell flashes. I should have said also that in the models which starts with a fair bit of hydrogen, roughly half of that hydrogen already burned before reaching that steady helium looping. Not too much more of the hydrogen was burned during the looping, much of it then burned later on, in shell flashes and steadily, with flashes going back and forth between the hydrogen and helium. Whatever you can imagine, it all happened.

Demarque: Do you really expect to see planetaries in large numbers in globular clusters? The only one that is known according to Peimbert has a mass in the nebula of the order of 1 hundredth of a solar mass. Now if horizontal branch stars have masses of the order of 6/10ths of a solar mass, they don't have much to eject, do they, and maybe they just bypass this phase; maybe one should expect to see them only in the old disc population where the masses are higher.

Salpeter: I quite agree that that's at least a possibility and even if Cahn and Kaler are right for population I, one has no way of knowing whether one should or should not expect them in globular clusters. It may even be that planetary nebulae are rare in globular clusters because of the lack of H shells but these ultraviolet stars are present. Incidently, I forgot to mention another indirect datum on this; Frank Kerr just earlier this week mentioned that searches for neutral and for ionised hydrogen in globular clusters have been made. Nothing has been found down to quite a low level. On the other hand, if anything I've said here is at all correct, that is not at all surprising because these ultraviolet stars, even after the nebula has become optically thin, still keep on driving the nebula out and so the velocities one gets (especially if some dust is present in the nebulae) are up to about 100 km s^{-1} and will have no difficulty escaping, so one would not predict theoretically that there should be hydrogen visible in globular clusters.

Faulkner: Are you now rejecting the correlation between observed nuclear properties as obtained from the Zanstra temperature and the Shklovski luminosities and the radius of the nebular shell which was the basis for assigning a several times 10^4 yr time scale on planetary nuclear evolution, in favour of this 10 times shorter scale?

Salpeter: Well I'm not quite sure whether I am or not, I'm at least foreshadowing the rejection and that was in part what my last slide was for. The selection effect might mimic the horseshoe curve which might make you think that you already are going down into the white dwarf region but still at a redder colour. The other feature that looks a little bit like a hook are these loopings, that's why I emphasize the loopings. It might also be that some early planetary nebulae are still in a transient stage and that for some reason it's this part of the helium burning loops that are most prominent.

Iben: Could you elucidate once more the manner in which you make the hydrogen invisible? Before doing so, let me give you some numbers: if mass is lost from stars in the giant phase before becoming horizontal branch stars, to the tune of about 2/10ths $M_\odot$ per star and if the ejection velocities are larger than the escape velocity from the cluster, then you might expect something like a solar mass of hydrogen in the vicinity of the 10 parsec confines of the cluster. Now Kerr's limit is something like a solar mass or less, of neutral hydrogen and the limits on ionised hydrogen are something like 15 $M_\odot$ and ultraviolet radiation will certainly turn that neutral hydrogen into ionised H; I think there's enough ultraviolet from the very blue horizontal branch stars to do the job.

Salpeter: Assume that (unlike what Demarque said) there are planetary nebulae in globular clusters and that they provide the main mass loss: Even then these planetary nebulae (the central stars themselves and the ultraviolet stars that they become) are enough so that the hydrogen that escapes from the planetary nebula never even recombines and gets speeded up from 20 km s^{-1} up to about 100 km s^{-1} as it goes out. So therefore you never would see any neutral hydrogen at all, and you would only have maybe a dozen or so of those shells still within the confines of the globular cluster at any one time. ($\ll 15\ M_\odot$ in ionised gas).

Iben: I see; do you make the mass that comes off the planetary invisible?

Salpeter: Immediately, yes.

Webster: When deriving this evolutionary track, Harman and Seaton calculated the optical depth of the nebulae to radiation past the Lyman limit and found that the boundary between optically thick and thin nebulae was on the low temperature edge of the track. I think it is unlikely, for this reason, that the high temperature hook is the particular optical depth selection effect that you are suggesting.

Salpeter: Well, it may not be possible; but on the other hand the stage where the luminosity of the star ceases to matter for the nebula luminosity depends on the luminosity of the star and these excursions in luminosity are pretty big. I really haven't looked at the effect quantitatively.

PART V

ERUPTIVE AND EXPLOSIVE VARIABLES

EXPLOSIVE AND ERUPTIVE STARS

E. SCHATZMAN

DAF, Observatoire de Meudon, 92190 Meudon, France

Abstract. A summary of the main results and problems concerning explosive and eruptive stars is given. The way in which the observations are related to the internal structure problems is briefly explained. Suggestions for further work are made: (1) dynamo theory, supersonic turbulence, stability of the envelope of T Tau stars; (2) rate of exchange of matter in close binary as the final clue to the thermal runaway theory of novae.

1. Aims and Limits of the Review

Explosive and eruptive stars are two very different classes of non stable stars. This has been shown a long time ago by Schatzman (1955) and by Ledoux (1957) (Table I),

TABLE I

The ratio (E/Lt) of the energy available during outbursts and in between outbursts

Class	$\Delta L/L$	E/Lt
Flare stars	5–100	$\simeq 0.01$
T Tauri stars	10– 20	1 ?
U Geminorum stars	30– 50	3 to 5
Recurrent novae	very large	1 to 1/2
Classical novae		1 to 1/2 ?

by comparing the energy available during the explosive or eruptive phase E, and the energy available Lt in between two such phases. The comparison is clearly in favour of a nuclear process for the explosive stars, and in favour of a less concentrated storage of energy for the eruptive stars. In fact, it seems quite reasonable to assume that the eruptive processes, similar to the solar eruptive processes, are due to some kind of plasma instability.

As far as we are concerned here, we do not intend to deal with all aspects of the eruptive and explosive stars, but only with the properties which derive from the internal constitution of these objects and the instability mechanisms.

2. Eruptive Stars

2.1. Properties and Classification

In the absence of a complete understanding of the eruptive stars, it is necessary to rely first on the empirical results concerning this class of stars. It is generally accepted, since the work of Haro (1968), Herbig (1962) and Ambarzoumian and Mirzoyan (1971), that the Herbig-Haro objects, the T Tauri stars and the UV Ceti stars represent an evolutionary sequence. The continuity of the spectroscopic properties is the basis of the idea of the physical continuity of the sequence.

However, the variability of the T Tau stars and that of the UV Ceti stars have a completely different behaviour. The nature of the variability of the UV Ceti stars seems to be well established since Lovell *et al.* (1963), following a suggestion by Schatzman (1959), proved the association of radio bursts (similar to the type III and type II solar radio bursts) with the optical flare stars. The association suggests very strongly that an electromagnetic process, very similar to the process observed at the surface of the Sun, takes place at the surface of the flare stars. This can be interpreted as due to the presence of a deep convective zone, in which a dynamo mechanism generates a system of magnetic spots, which, in turn generate the instability responsible both for the optical flare and for the fast particles which emit or stimulate the radio emission. Observation of the rotation of a dark spot at the surface of d Me stars by Torres and Ferraz Mello (1973) is very much in favour of such an interpretation. Gershberg (1970a, 1970b) claims that the observations are consistent with the assumption that the star has a rotation period of several hours and that the strong flares occur only in some active region of the star surface.

The question of a period of activity, similar to the solar cycle has been raised. The IAU (1970) report of the working group on flare stars mentions that no conclusions have been reached as yet.

The analysis by Kuhi (1964) of the spectrum of the T Tauri stars was very successful in explaining correctly the excitation and the line profile of the emission spectrum. However, his velocity field implies that matter leaves the star without having the escape velocity but does not fall back on the star. A recent model of Prentice (1973) suggests that matter goes up, under the action of buoyancy force, in needle like supersonic eddies which are finally decelerated and then form a rarefied descending atmosphere.

These models account for the T Tau star 'mass ejection' phenomenon, without any net mass loss for the star. Prentice's model presents the interest of building up a dense outer non-turbulent shell. The corresponding inversion density strongly suggests the presence of a Taylor instability. This has not been studied by Prentice, but we can estimate the time scale of the instability $\sigma^2 = gk$, by assuming that k^{-1} is of the order of the scale height of the supersonic turbulent region. This gives a time scale for the instability of the order of a day. This is clearly not of the order of the time scale of the light variations of T Tauri stars, as given by Herbig (1962).

The suggestion made here is that the T Tauri activity is not a blend of flares of varying intensity but is related to the building of the outer layers. It is not possible, for the time being, to describe more precisely a model of the instability of the T Tauri stars.

2.2. Evolution and age

The model which relates stellar rotation, age, and electromagnetic activity, is now well accepted. However, the present state of the dynamo theory does not give the possibility of a precise description of the observed phenomena. The dynamo theory, in its various forms (Lerche 1971a, 1971b, and Steinbeck and Krause, 1969) gives a good description of the behaviour of the average magnetic field, including both the effects of helicity and rotation. However, the description of the properties of the rms fluctuations of the magnetic field, $\langle(|\delta B|)^2\rangle^{1/2}$ is still insufficient to lead, for example, to a real understanding of the production of the sun spots. It is likely that the cyclic activity of a star is related to the thickness of the convective zone. A fully convective star is unlikely to behave like a star with a convective zone of finite thickness. In the same way, the rate of change of the activity with the angular velocity is, for the time being a purely empirical fact, without any quantitative estimate, but for the trivial statement that the activity decreases with the angular velocity. This can be considered as being confirmed by the observations of flash stars in clusters. According to Haro (1968), flare stars are less massive and colder in old clusters. As shown by Gershberg and Pikel'ner (1972), the properties of flare stars fit well in the whole picture of stellar evolution, activity and stellar structure.

The amount of energy which goes into mass ejection, and the magnetic field determine the rate of exchange of angular momentum. Gershberg and Pikel'ner (1972) give estimates of the relevant quantities. Unsöld (1957) suggested that a substantial part of the galactic cosmic ray flux has its origin in stellar flares. Hudson *et al.* (1971) suggested the existence of a class of unresolved low luminosity X-ray sources to account for the observations. Edwards (1971) supports the suggestion of Hudson *et al.* but shows that the energy radiated in soft X-rays is probably of the order of 1% of the optical flare energy, more or less in agreement with the value of 10% quoted by Thomas and Teske (1971) for the solar flares.

The frequency of the optical flares as a function of their magnitude has been given by Kunkel (1973). It gives the possibility of estimating more precisely the amount of energy going into the flares, which is altogether a small fraction of the luminosity of the star. As a consequence, the rate of evolution of flare stars is not appreciably changed when including the energy of the flares. This is confirmed by the analysis of Gurzadyan (1971).

Further study of flare stars present a great interest both from the physical and the astrophysical points of view:

(1) It can lead to a better and quantitative understanding of the dynamo mechanism.

(2) It can lead to a better and quantitative understanding of the relation between age and rotation.

(3) Other properties are more directly related to the physics of the surface phenomena (X-rays, cosmic rays, radio emission) and are important both in themselves (e.g. accelerating mechanisms) and for the evolutionary processes.

3. Explosive Stars

3.1. Models of novae and U Geminorum stars

The suggestion by Schatzman (1956, 1958) that all prenovae and U Geminorum stars are double stars was proved by Kraft (1964). Since that time the greatest advance in the subject was probably the analysis, by Saslaw (1968) of the thermal runaway due to accretion. The white dwarf, in the double star, accretes matter shed from the larger star through the inner Lagrange point.

The thermal runaway depends, for a given mass, both from the internal temperature of the white dwarf and from the rate of accretion. Similar models have been studied by Giannone and Weigert (1967) and by Rose (1968), but failed to yield the typical energy of nova DQ Herculis 1934. By including the effect of an accretion shock Starrfield (1971a, 1971b), succeeded in obtaining the necessary energy. The objections of Schatzman (1965) to the thermal runaway were valid only for a very slow rise of the rate of energy generation and are not valid in the case of fast accretion.

Further work by Starrfield *et al.* (1972), using a more detailed hydrodynamic code, has led to a better agreement. Their result is sustained by the production of ^{13}C, ^{17}O. Further confirmation can be found in the analysis of the isotopic production in hot CNO-Ne cycle by Audouze *et al.* (1973). Altogether, the thermal runaway seems now the well established mechanism of the nova outburst. Weaver (1973) has shown that the model leads to a complete understanding of Nova Aql 1918.

The interest of the thermal runaway is that it can produce an increase in luminosity without any mass loss. This seems to be the case of the U Gem stars. The connection between the theory of thermal runaway and the impulsive generation of non radial oscillations is well established by the observations of Warner (1974).

4. Conclusions

As far as such general considerations are useful, we may notice that the theory of explosive variables is much more advanced than the theory of eruptive variables. The reason is that the explosions depend on thermonuclear processes and transport processes which are fairly well understood and leaves out completely the problem of the accretion rate, which is just kept as a parameter. On the other hand, the mechanism of eruptive variables is due to turbulent processes which remain presently one of the unresolved problems of physics.

References

Ambarzoumian, V. A. and Mirzoyan, L. V.: 1971, *IAU Colloq.* **15**, *Veröffentlichungen der Remeis-Sternwarte Bamberg*, No. 100, p. 98.
Audouze, J., Truran, J. W., and Zimmermann, B. A.: 1973, Orange AID preprint 315.
Edwards, P. J.: 1971, *Nature Phys. Sci.* **234**, 75.
Gershberg, R. E.: 1970a, *Flares of Red Dwarf Stars*, Moscow, transl. into English by D. J. Mullan (Armagh Observatory, Armagh, Northern Ireland).
Gershberg, R. E.: 1970b, in A. Slettebak (ed.), 'Stellar Rotation', *IAU Colloq.* **4**, 249.
Gershberg, R. E. and Pikel'ner, S. B.: 1972, *Comments Astrophys. Space Phys.* **4**, 113.
Giannone, P. and Weigert, A.: 1967, *Z. Astrophys.* **67**, 41.
Gurzadyan, G. A.: 1971, *Bol. Obs. Tonantzintla Tacubaya* **6**, 39.
Haro, G.: 1968, *Stars and Stellar Systems* **7**, 141.
Herbig, G. H.: 1962, in Z. Kopal (ed.), *Advances in Astronomy and Astrophysics*, Vol. 1, 47, Academic Press, London.

Hudson, H. S., Peterson, L. E., and Schwartz, D. A.: 1971, *Nature* **230**, 177.
IAU: 1970, *Transactions of the IAU*, Vol. XIVA, Report of the Working Group on Flare Stars, p. 297.
Kraft, R.: 1964, *Astrophys. J.* **139**, 457.
Kuhi, L.: 1964, *Astrophys. J.* **140**, 1409.
Kunkel, W. E.: 1973, *Astrophys. J. Suppl.* **25**, 1.
Ledoux, P.: 1957, in G. H. Herbig (ed.), *Non Stable Stars*, Cambridge University Press, p. 181.
Lerche, I.: 1971a, *Astrophys. J.* **166**, 627.
Lerche, I.: 1971b, *Astrophys. J.* **166**, 639.
Lovell, P., Whipple, F. L., and Solomon, H, L.: 1963, *Nature* **198**, 228.
Prentice, A. J. R.: 1973, *Astron. Astrophys.* **27**, 237.
Rose, W. K.: 1968, *Astrophys. J.* **152**, 245.
Saslaw, W. C.: 1968, *Monthly Notices Roy. Astron. Soc.* **138**, 337.
Schatzman, E.: 1955, Principes Fondamentaux de Classification Stellaire, Colloque international du CNRS No. 55, Publications du CNRS, Paris, p. 175.
Schatzman, E.: 1956, *Non Stable Stars*, Symposium of the Acad. Sci. of U.S.S.R., Publishing House of the Acad. Sci., Erevan, p. 141.
Schatzman, E.: 1958, *Ann. Astrophys.* **21**, 1.
Schatzman, E.: 1959, in R. N. Bracewell (ed.), 'Paris Symposium on Radio Astronomy', *IAU Symp.* **9**, 552.
Schatzman, E.: 1965, *Stars and Stellar Systems* **8**, 327.
Starrfield, S.: 1971a, *Monthly Notices Roy. Astron. Soc.* **152**, 307.
Starrfield, S.: 1971b, *Monthly Notices Roy. Astron. Soc.* **155**, 129.
Starrfield, S., Truran, J. W., Sparks, W. M., and Kutter, G. S.: 1972, *Astrophys. J.* **176**, 169.
Steenbeck, von M. and Krause, F.: 1969, *Astron. Nachr.* **291**, 49.
Thomas, R. J. and Teske, R. G.: 1971, *Solar Phys.* **16**, 431.
Torres, C. A. O. and Ferraz Mello, S.: 1973, *Astron. Astrophys.* **27**, 231.
Unsöld, A.: 1957, in H. C. van de Hulst (ed.), 'Radio Astronomy', *IAU Symp.* **4**, 238.
Warner, B.: 1974, this volume, p. 125.
Weaver, H.: 1973, private communication.

ON THE BINARY NATURE OF THE SLOW NOVA, RR TELESCOPII

B. L. WEBSTER

Royal Greenwich Observatory, Hailsham, Sussex, England

Abstract. Those novae known to be binaries generally have orbital periods of the order of hours, exceptions being the atypical recurrent novae T CrB and RS Oph, which have giant companions and probably much longer periods. Since the orbital period in a semi-detached system relates to the mechanism of current mass exchange and also to the extent to which the primary evolved before mass exchange took place at an earlier stage, it is of interest to see if any classical novae are in more widely separated systems.

This communication concerns the star RR Telescopii, which has all the characteristics of a slow nova – a range in amplitude greater than seven magnitudes, a spectral type at maximum of F, and a decline through a nebular stage of increasing ionization level (e.g. Thackeray, 1955). RR Tel was seen as a variable before outburst, but little is known about this variable apart from its period of 387 days, although doubts have been expressed about its being a red variable (Payne-Gaposchkin, 1957). Dr Thackeray has Radcliffe spectrograms of RR Tel from soon after maximum to the present. On some of the more recent of these, bands of TiO have become visible, presumably as the hot star has faded, and Dr Thackeray and I interpret these as meaning that the original variable is still there and is an M giant. Thus RR Tel is in a binary system containing a giant M star and a hot companion and such a system cannot have an orbital period of hours like the classical novae mentioned above.

In 1972, Dr Glass and I examined RR Tel in the infrared between 1.2 and 20 μ. The energy distribution does not resemble the cool star that might be expected, but is exactly like that of free-free radiation over the whole wavelength range. The puzzle is that the infrared is two orders of magnitude stronger than we would predict from the optical spectrum for free-free radiation.

References

Payne-Gaposchkin, C.: 1957, in J. H. Oort, M. G. J. Minnaert, and H. C. van de Hulst (eds.), *The Galactic Novae*, North-Holland Publishing Company, Amsterdam, p. 157.
Thackeray, A. D.: 1955, *Monthly Notices Roy. Astron. Soc.* **115**, 236.

DISCUSSION

Schatzman: I am well aware that this description that I gave does not fit to all stars which are called novae. I would rather say that as far as novae are concerned, in the present situation there are processes which we understand and those we do not understand, and I am afraid I have spoken only of what *I* understand. If I could give other examples, there are difficulties for example like nova Del which has been a very extraordinary object and many others (you mentioned the case of this nova where the double star has appeared). What I think is possible is that we call novae a number of objects which are *not* exactly of the same physical state and the differences show up in the spectrum and in the light curve. I have concentrated, myself, either on the mechanism by which the outbursts can be produced in some cases rather than on the way one can derive from the observation the process which has taken place.

NON-RADIAL PULSATIONS OF DWARF NOVAE

BRIAN WARNER

University of Cape Town, Rondebosch, South Africa

Abstract. The high speed photometric observations of dwarf novae during outbursts that led to the discovery (Warner and Robinson, 1972) of rapid pulsations have been extended to other objects. Pulsations are usually seen near outburst maximum; an exception is VW Hyi which showed oscillations on December 25, 1972, when it was well down the declining part of its light curve.

The periods of the oscillations are found to vary; in most cases a rapid change from one discrete mode to another is observed. The dwarf nova stays in one oscillation mode for about an hour, and changes mode over a time of about 10 min. In the case of the VW Hyi observation, the period changed continuously from 28 s to 34 s over a time of 4 h, during which the apparent brightness decreased by 40%.

The existence of many closely spaced oscillation modes suggests that the dwarf novae are oscillating in non-radial modes. This is supported by observations of the 71-s oscillations of DQ Her: during eclipse a phase shift is seen indicating an $l=2$, $m=2$ mode of oscillation. In UX UMa, the 29-s oscillations show an opposite phase shift from that in DQ Her and indicate an $l=2$, $m=-2$ mode.

A period-luminosity relationship can be deduced for the oscillations in dwarf novae, and this is in good agreement with recent calculations on quadrupole oscillations of hot white dwarfs made by Osaki and Hansen (1973).

Pulsations have so far been detected in the following objects: Z Cam (17 s), SY Cnc (24.6), CN Ori (25), KT Per (27), UX UMa (29), VW Hyi (28–34), CD −42°14462 (29 and 30), AH Her (31).

References

Osaki, Y. and Hansen, C. J.: 1973, *Astrophys. J.* **185**, 277.
Warner, B. and Robinson, E. L.: 1972, *Nature* **239**, 2.

TESTS OF THE BINARY HYPOTHESIS OF NOVAE THROUGH NOVA NEBULAE OBSERVATIONS

J. B. HUTCHINGS

Dominion Astrophysical Observatory, Victoria, B.C., Canada

Abstract. I should like to follow up Dr Schatzman's talk with a few words on models of Nova nebulae. Geometrical models for novae Aql 1921 and Her 1934 have been derived by Mustel and Boyarchuk (1970), for novae Del 1967, Vul 1968 and Ser 1970 by Hutchings (1972), for Aql 1921 by Weaver (1974) and for Del 1967 by Malakpur (1973). Very different techniques and lines of reasoning have been used by all these people, but the models derived all have very strong similarities which should lead us to take them seriously in considering the nature of the nova mechanism.

It appears that nova remnants are characterized by two polar cones or blobs of similar but opposite velocity and similar density, moving along the axis of rotation of the system. In addition, one or more rings of matter may exist symmetrically about the equatorial plane, with a recession velocity and density which, in general, are different from the polar matter. The polar blobs appear to be associated with the matter giving rise to the 'principal' spectrum in the bright phases of the nova.

Three possible main causes have been suggested for the formation of such remnants, all connected with the initial outburst: (1) Weaver has proposed that an initially spherically symmetrical expansion is impeded and deflected by matter circulating in an accretion ring around the degenerate star. (2) Schatzman has suggested that the presence of the ring may favour an initial outburst preferentially in the polar regions. (3) Hutchings has suggested that physical interference of an initially spherical expansion may occur by a secondary component whose size is a large fraction of the separation of the two stars.

I feel that these points need to be considered by theoreticians in proposing detailed mechanisms of the nova outburst.

References

Hutchings, J. B.: 1972, *Monthly Notices Roy. Astron. Soc.* **158**, 177.
Malakpur, I.: 1973, *Astron. Astrophys.* **24**, 125.
Mustel, E. R. and Boyarchuk, A. A.: 1970, *Astrophys. Space Sci.* **6**, 183.
Weaver, H.: 1974, in G. Contopoulos (ed.), *Highlights of Astronomy*, **3**, XVth General Assembly of the IAU, D. Reidel Publishing Company, Dordrecht, Holland, p. 423.

PROPOSED SEARCH FOR FAST VARIABLES

J. G. DUTHIE, B. RENAUD, and M. P. SAVEDOFF
University of Rochester, Rochester, N.Y. 14627, U.S.A.

We propose the examination of trailed photographs as a medium of search for fast variable stars. Diurnal rotation and seeing permit access to the period range from 0.1^s to 10^s for a resolution of $1''$, exposures of one minute and a plate field of $30'$, while slower or faster periods can be studied by mechanically trailing at the appropriate rate. This permits examination of a wide period band over a field of perhaps 0.1 square degree in a minute's time, while photoelectric devices now available permit only star by star examination. Thus we can contemplate examination of an appreciable fraction of the sky with moderate telescope time required.

Fourier transform processing promises an efficient technique for analogue image processing. As a demonstration, we have prepared an artificial field consisting of uniform trails of some minute's duration (non-variable) upon which we superimposed some twenty exposures of three seconds duration each displaced identically in the diurnal direction by some $12''$ (variables). If this plate replaces the transmission grating of a spectrograph, monochromatic illumination of the entrance aperture will produce in the image plane the Fourier transform of the plate (Fourier Image); each set of twenty exposures acts as a narrow grating and produces a grating spectrum, all superimposed, as the image is independent of vertical and horizontal translations of the grating. Dominant in this image is the transform (diffraction pattern) of the plate grain and the aperture. A subsequent Fourier transform, produces an image of the plate (grating) we call the Processed Image which may be modified by inserting spatial filters in the Fourier Image plane to enhance the visibility of the variable stars with respect to the non-variable stars, grains, etc. Essentially, if the spatial filter is the transform of the 'target' signal, then the Processed Image is the cross-correlation of the original image with the target signal; non-cross-correlated signals are strongly suppressed. In particular, a ten slot spatial filter, with period P, matched to our plate, width of slots $P/10$, completely suppressed non-variable images (bars) while the twenty exposure sequences are plainly visible. The random cross-correlations with the background noise provides the limit of detectability for faint strongly modulated variable stars. When the target period is mismatched by 10% both the bars and the variables disappear.

In practice, 11th magnitude stars have been easily detected on our artificial fields prepared on Eastman Kodak 103F emulsion at the 60 cm Cassegrain of the C. E. K. Mees Observatory. We anticipate reaching the same magnitude on trailed exposures in the presence of fog, and perhaps 15th magnitude if the fog background could be reduced by a happier choice of emulsion and processing techniques. This technique is unique in allowing full use of the high detection quantum efficiency of photographic emulsions at low exposure levels.

We have been unable to completely analyze the anticipated signal to noise properties of the system in the presence of non-linear photographic effects. Clearly, as in ordinary image recognition we seek to recognize a variation of the plate grain density with respect to the background. Unlike a direct photograph, our image, while containing the same number of grains, is distributed over the length of the trail; the noise background is increased relative to a direct photograph by $(Bs/d)^{1/2}$ where B is the ratio of mean grain densities, s is the trail length and d, the width. For large amplitudes, $B \sim 1$ while for an equatorial one minute exposure, s is 900" while $d \sim 1"$, hence the noise is some 30 times larger than on guided images for the same exposure time. This implies a loss of 7.0 magnitudes as compared to direct exposure and is the basis for the estimate above. The actual number of grains dominates the noise for fog-free case, and we have assumed above that the number of grains is one-tenth the number of photoelectrons counted with modern equipment for a 15th magnitude star.

The concepts introduced here arose from suggestions by Charman (1968) on the usefulness of the first Fourier transform in searching for variables. Photoelectric searches using tuned recorders by Duthie et al. (1968) and fast tape recording by Horowitz et al. (1971) could only search a small area (pin-hole) at a time, while the photographic techniques pioneered by Wampler and Miller (1969) and applied by Chiu et al. (1971), while more sensitive than ours and panoramic, require sufficient period information to ensure phase coherency over the entire set of exposures. The study of trailed star images with the use of image processing permits efficient search for fast variable stars over large portions of the sky. Although the sensitivity of the search is less than that for direct photography, the opportunity to examine many images efficiently promises a statistical base for discussions of the numbers and duration of fast stellar variability.

References

Charman, W. N.: 1968, *Appl. Opt.* **7**, 2431.
Chiu, H., Lynds, R., and Maran, S.: 1971, *Publ. Astron. Soc. Pacific* **82**, 660.
Duthie, J. G., Sturch, C., and Hafner, E. M.: 1968, *Science* **160**, 1104.
Horowitz, P., Papaliolios, C., and Carlton, N.: 1971, *Astrophys. J. Letters* **163**, L5.
Wampler, E. J. and Miller, J. S.: 1969, *Nature* **221**, 1037.

PART VI

INSTABILITY MECHANISMS

A. NON-RADIAL PULSATIONS – MAGNETIC FIELDS

NON-RADIAL OSCILLATIONS

P. LEDOUX

Institut d'Astrophysique, Université de Liège, Belgium

Abstract. The problem of the adiabatic non-radial oscillations of spherical stars is reviewed and results recalled for a variety of models. The anomalous behaviour of the eigenfunctions for highly condensed models is related to the apparent mobile singularities depending on the eigenvalues. Tables of Q-values are provided to facilitate possible applications to variable stars.

In the case of the gravity modes, the existence of multiple spectra some stable (g^+ modes) some unstable (g^- modes) if superadiabatic and subadiabatic regions alternate is discussed and the interest of further investigations underlined.

As far as vibrational stability (effects of the non adiabatic terms) is concerned, a general expression is given for the 'damping coefficient'. The attention is drawn to the possibility for g^+ modes of becoming vibrationally unstable under the effect of various factors and in various models, including the Sun where this was advocated as a possibility of relieving the neutrinos difficulty.

Finally the present status of the most obvious candidates among variable stars for non-radial oscillations, the β Canis Majoris stars and the rapid blue variables (white dwarfs) is briefly reviewed.

1. Introduction

I shall devote this introductory lecture mainly to linear non-radial oscillations of purely spherical stars i.e. devoid of rotation or magnetic fields or of tides which, in general, would imply deviations from spherical symmetry. The effects of some of these factors will be treated in other lectures or communications during this session and I shall limit myself to a few comments on some of the simplest aspects of their influence.

There has been lately quite a renewal of interest in the response of stars to non-radial perturbations aroused either by attempts at interpreting some types of variable stars like the β Canis Majoris stars and the new white dwarf variables, or by phenomena in the external layers of the Sun like the 5-min oscillation discovered by Leighton, or by the hope to add somewhat to our knowledge of convection and its penetration in nearby convectively stable zones, or by the desire to explore some new aspects of stellar stability which may be of great importance for the evolution of the star. On the other hand, one must expect that such non-radial motions should be easily excited in a variety of close double stars with eccentric orbits and it is likely that, with the extraordinary progress in observational techniques, these should become observable and be identified as such pretty soon. Finally there is direct evidence in novae, perhaps even in planetary nebulae, for the presence of non-radial velocity fields.

There exist general review articles centered on the stellar case (Ledoux and Walraven, 1958; Ledoux, 1969) but the problem is one of importance in many branches of geophysics as well (meteorology, oceanography and also tidal or free oscillations of the Earth) and the literature in this field is quite extensive, a few general accounts being also available (cf. Eckart, 1960; Tolstoy, 1963; Bolt and Derr, 1969). One should also mention the books by Chandrasekhar (1961, 1969) where many related problems are treated and methods presented which are of general interest in this context.

2. General Definitions and Equations

The simplest way to introduce the indispensable definitions and notations is still to write the general linearized equations derived from the conservation of mass, momentum and energy and from Poisson's equation. If the initial non-perturbed configuration is strictly in hydrostatic equilibrium and assuming a time dependence for all perturbations of the form $e^{i\sigma t}$, these equations can be written with fairly standard notations (cf. Ledoux and Walraven, 1958)

$$\frac{\delta\varrho}{\varrho} = \frac{\varrho'}{\varrho} + \frac{\delta r}{\varrho}\frac{d\varrho}{dr} = -\text{div }\delta\mathbf{r} \tag{1a}$$

$$\sigma^2\delta\mathbf{r} - \text{grad }\Phi' + \frac{\varrho'}{\varrho^2}\text{grad }p - \frac{1}{\varrho}\text{grad }p' = -\frac{i\sigma}{\varrho}\text{div }\mathbf{P}(\delta\mathbf{r}) \tag{1b}$$

or

$$\sigma^2\delta\mathbf{r} - \text{grad}\left(\phi' + \frac{p'}{\varrho}\right) + \frac{\mathbf{r}}{r}\mathscr{A}\frac{\Gamma_1 p}{\varrho}\text{div }\delta\mathbf{r} =$$
$$= \frac{\Gamma_3 - 1}{i\sigma}\frac{1}{\varrho}\text{grad}\varrho\left(\varepsilon - \frac{1}{\varrho}\text{div}\mathbf{F}\right)' - \frac{i\sigma}{\varrho}\text{div }\mathbf{P}(\delta\mathbf{r}) \tag{1b'}$$

$$\delta p - \frac{\Gamma_1 p}{\varrho}\delta\varrho = p' + \delta r\frac{dp}{dr} - \frac{\Gamma_1 p}{\varrho}\left(\varrho' + \delta r\frac{d\varrho}{dr}\right) = \frac{\Gamma_3 - 1}{i\sigma}\varrho\left(\varepsilon - \frac{1}{\varrho}\text{div}\mathbf{F}\right)' \tag{1c}$$

which can also be written

$$p' - \Gamma_1 p\left(\frac{\varrho'}{\varrho} + \mathscr{A}\delta\mathbf{r}\right) = \frac{\Gamma_3 - 1}{i\sigma}\varrho\left(\varepsilon - \frac{1}{\varrho}\text{div}\mathbf{F}\right)' \tag{1c'}$$

or in terms of the variables T and p

$$\delta T - \frac{\Gamma_2 - 1}{\Gamma_2}\frac{T}{p}\delta p = T' + \delta r\frac{dT}{dr} - \frac{\Gamma_2 - 1}{\Gamma_2}\frac{T}{p}\left(p' + \delta r\frac{dp}{dr}\right) =$$
$$= \frac{1}{i\sigma C_p}\left(\varepsilon - \frac{1}{\varrho}\text{div}\mathbf{F}\right)' \tag{1c''}$$

or

$$T' - T\left(\frac{\Gamma_2 - 1}{\Gamma_2}\frac{p'}{p} + \delta r\mathscr{S}\right) = \frac{1}{i\sigma C_p}\left(\varepsilon - \frac{1}{\varrho}\text{div}\mathbf{F}\right)' \tag{1c'''}$$

$$\nabla^2\Phi' = 4\pi G\varrho', \tag{1d}$$

where a prime denotes an Eulerian perturbation and δ a Lagrangian one. The right-hand members of (1b) and (1c) represent non-conservative terms related to the generation of energy ε and its flux (**F**) and to the effects of the viscous stresses **P**(δ**r**). In general these terms are much smaller than the others in (1b) and (1c) and neglecting them yields the *adiabatic approximation*.

The quantity $\mathscr{A}$ defined by

$$\mathscr{A} = \frac{1}{\varrho}\frac{d\varrho}{dr} - \frac{1}{\Gamma_1 p}\frac{dp}{dr} \tag{2}$$

is the argument of the criterion for convection which develops or not depending on whether $\mathscr{A} > 0$ or $\mathscr{A} < 0$. $\mathscr{A}$ is directly related to the Brunt-Väisälä frequency N by

$$N^2 = -g\mathscr{A},$$

where g is the gravity. The quantity

$$\mathscr{S} = \frac{\Gamma_2 - 1}{\Gamma_2}\frac{1}{p}\frac{dp}{dr} - \frac{1}{T}\frac{dT}{dr} \tag{2'}$$

is the argument of Schwarzschild criterion which, for constant chemical composition (mean molecular weight $\bar{\mu} = $ ct.), is sufficient to determine whether convection develops ($\mathscr{S} > 0$) or not ($\mathscr{S} < 0$). For a mixture of perfect gas and radiation, one has

$$\mathscr{A} = \frac{1}{\bar{\mu}}\frac{d\bar{\mu}}{dr} + \frac{4 - 3\beta}{\beta}\mathscr{S}, \tag{2''}$$

where β is the ratio of the gas to the total pressure. Thus, if $\bar{\mu}$ decreases upwards, convection may not appear even if $\mathscr{S} > 0$.

We shall limit ourselves first to the adiabatic case (r.h.m. of (1b) and (1c) identically zero) and note that, in spherical polar coordinates (r, θ, φ), the radial component δr of the displacement as well as p', ϱ' and Φ' can be factorized in the form

$$f'(r, \theta, \varphi) = f'(r)\, P_l^m(\cos\theta)\, e^{im\varphi}, \quad -l \leq m \leq l$$

where $P_l^m(\cos\theta)$ is the associated Legendre polynomial of degree l and order m so that

$$\mathrm{div}\,\delta\mathbf{r} = \frac{1}{r^2}\frac{d}{dr}(r^2\delta r) - \frac{l(l+1)}{\sigma^2 r^2}\left(\Phi' + \frac{p'}{\varrho}\right) \tag{3}$$

and

$$\nabla^2\Phi' = \frac{1}{r^2}\frac{d}{dr}\left(r^2\frac{d\Phi'}{dr}\right) - \frac{l(l+1)}{r^2}\Phi' = 4\pi G\varrho'. \tag{4}$$

System (1) where we take only the r-component of Equation (1b'), i.e. here

$$\sigma^2\delta r = \frac{d}{dr}\left(\Phi' + \frac{p'}{\varrho}\right) - \mathscr{A}\frac{\Gamma_1 p}{\varrho}\,\mathrm{div}\,\delta\mathbf{r} \tag{5}$$

becomes then an ordinary differential system of the fourth order in the space variable r for $\delta r, p', \varrho', \Phi'$. Once its solution is known, the two other components of the equation of motion

$$\sigma^2 r \delta\theta = \left(\Phi' + \frac{p'}{\varrho}\right) \frac{1}{r} \frac{\partial P_l^m}{\partial \theta} e^{im\varphi} \tag{6}$$

$$\sigma^2 r \sin\theta \, \delta\varphi = \left(\Phi' + \frac{p'}{\varrho}\right) \mathrm{im} \frac{P_l^m}{r \sin\theta} e^{im\varphi} \tag{7}$$

serve to define the horizontal components of the displacements.

If we associate with the 4th order problem, the four natural boundary conditions:
δr: finite in $r=0$ (in fact $\delta r \propto r^{l-1}$ and $p', \varrho', \Phi' \propto r^l$ as $r \to 0$)
$\delta p = p' + \delta r (dp/dr) = 0$ in $r = R$
$(\Phi'_i)_R = (\Phi'_e)_R = C/R^{l+1}$ (continuity of potential across $r = R$)
$(d\Phi'_i/dr)_R + (l(l+1)/R)(\Phi'_i)_R = -(4\pi G \varrho \delta r)_R$ ($=0$ if $\varrho_R = 0$)
 (continuity of gravitational force across R),
we are left with a self-adjoint 4th order eigenvalue problem whose solutions, orthogonal to each other

$$\int_0^R \delta \mathbf{r}_i \cdot \delta \mathbf{r}_k^* \varrho \, d\mathscr{V} = 0 \quad i \neq k \tag{8}$$

define the stationary modes of non-radial oscillations.

As can be verified immediately from (1a, c, d), (5), (3) and (4) and the boundary conditions, this eigenvalue problem does not depend explicitly of m so that, for a given l, there are $(2l+1)$ solutions (corresponding to the various possible values of m in $-l \leq m \leq l$) associated with each eigenvalue. Fields of forces devoid of spherical symmetry can remove this $(2l+1)$-fold degeneracy either totally (for instance: rotation) or partially (for instance: magnetic field, tides).

One must also be aware that $\sigma^2 = 0$ is a highly degenerate trivial eigenvalue of the spherical configuration corresponding to a displacement normal to gravity with all corresponding $\varrho', p', \Phi' = 0$, which may give rise to significant eigenvalues (for instance, corresponding to toroidal oscillations) in presence of fields of forces of the type considered above so that the total number of distinct modes for a given l may become larger than $(2l+1)$ (cf. for instance Perdang, 1968, for a general group-theoretical discussion of the question).

Apart from this question of degeneracy, the 4th order eigenvalue problem is rather peculiar of the type which has recently drawn a certain amount of attention from mathematicians (for problems related to ours, cf. for instance: Eisenfeld, 1968a, b; Weinberger, 1968) under the name of 'non-linear' eigenvalue problems because, contrarily to the case of the classical Sturm-Liouville problem, the eigenvalue, here $\lambda = \sigma^2$, enters non-linearly in the coefficients. This translates mathematically the fact that, in general, the spectrum of eigenvalues splits into distinct spectra with quite different physical meanings.

To illustrate the situation, let us consider the approximation obtained by neglecting Φ' which for $l>0$ is usually a fairly good approximation except perhaps for the lowest modes since Φ', solution of Poisson's equation, is an integral expression over the whole star decreasing when l and the number of nodes of ϱ' along the radius increase. This has been confirmed by numerical integrations (Robe, 1968) which furthermore show that the characteristics of the spectra are not affected. In that case, we are left with a second order differential problem which can be written

$$\frac{du}{dr}+\frac{1}{\Gamma_1 p}\frac{dp}{dr}u=\left[\frac{l(l+1)}{\sigma^2}-\frac{\varrho r^2}{\Gamma_1 p}\right]y \tag{9a}$$

$$\frac{dy}{dr}+y\mathscr{A}=\frac{1}{r^2}(\sigma^2+\mathscr{A}g)u \tag{9b}$$

with $u=r^2\,\delta r$ and $y=\dfrac{p'}{\varrho}$.

The 'non-linear' character of the eigenvalue problem is now apparent. However, as pointed out by Cowling (1941) and ignoring for the time being possible difficulties towards the extremities of the interval, this problem with the first two boundary conditions tends to a Sturm-Liouville problem for $|\sigma^2|$ either very large or very small.

Indeed, in the first case, eliminating u from (9a and b) and neglecting terms of the order of $1/\sigma^2$, we obtain

$$\frac{d^2 y}{dr^2}+\frac{dy}{dr}\left(\frac{2}{r}+\frac{1}{\varrho}\frac{d\varrho}{dr}\right)+y\left[\frac{\sigma^2\varrho}{\Gamma_1 p}+\frac{1}{r^2}\frac{d}{dr}(r^2\mathscr{A})-\frac{l(l+1)}{r^2}\right]=0. \tag{10}$$

The only subsisting term containing σ^2 being $\sigma^2\varrho/\Gamma_1 p$, the solutions must correspond to acoustical or pressure (p) modes associated with waves propagating with the velocity of sound $c=(\Gamma_1 p/\varrho)^{1/2}$. It can also be shown that, for all realistic values of the physical parameters, all these p modes have discrete positive eigenvalues, $\sigma_p^2>0$, with an accumulation point at $+\infty$. Thus no instability can arise through these modes. This is very different from the radial case ($l=0$) for which a very important dynamical instability enters through the fundamental mode if $\Gamma_1<4/3$. Nevertheless, as we shall see later, there are many close analogies with the radial modes which are also acoustical modes.

If σ^2 is sufficiently small, eliminating y between (9a and b) and neglecting terms of the order of σ^2 yields

$$\frac{d^2 u}{dr^2}+\frac{du}{dr}\frac{1}{\varrho}\frac{d\varrho}{dr}+u\left[-\frac{\mathscr{A}g\,l(l+1)}{\sigma^2}\frac{1}{r^2}-\frac{l(l+1)}{r^2}+\frac{d}{dr}\left(\frac{1}{\Gamma_1 p}\frac{dp}{dr}\right)\right]=0. \tag{11}$$

This is again of the type of a Sturm-Liouville problem for which the parameter $\lambda=1/\sigma^2$

will have an infinite discrete spectrum with a point of accumulation at infinity, i.e. σ^2 will take an infinite number of discrete values accumulating at zero.

If we assume $\mathscr{A} < 0$ everywhere (convective stability), these solutions correspond to gravity or g modes associated with waves propagating with a velocity of the order of $r\sigma^2/[l(l+1)(-\mathscr{A})g]^{1/2}$. In that case and provided $d/dr\,[(1/\Gamma_1 p)\,dp/dr]$ be negative everywhere, which is usually realized in realistic conditions, these σ^2 are all positive and the star is dynamically stable.

However, if $\mathscr{A} > 0$ everywhere, all the eigenvalues σ^2 become negative and dynamical instability sets in giving rise to convective motions inside the star. In other words, the only dynamical instability which may enter through non-radial perturbations is the type of violent internal motions known usually as convection. It can also be shown (Lebovitz, 1965, 1966) that $\mathscr{A} > 0$ in any finite interval however small implies the existence of negative σ_g^2 and convection in that region so that $\mathscr{A} < 0$ is really a necessary and sufficient condition for convective stability.

In the same paper, Cowling (1941) established also the existence of an intermediate mode which he called fundamental or f mode with normally no nodes in δr or ϱ' and with a positive eigenvalue σ_f^2 separating the σ_g^2 and σ_p^2.

All this has been confirmed and somewhat clarified on the basis of a variational principle by Chandrasekhar (1964) and Chandrasekhar and Lebovitz (1964) who have shown that the eigenvalues σ^2 of the general 4th order problem expressed in terms of integrals on the displacement $\delta \mathbf{r}$ are stationary when the latter tends to an eigensolution. In other words, each eigenvalue is an extremum but we don't know whether it is a maximum or a minimum. On the other hand, the discussion above suggests directly that asymptotically the σ_p^2, as eigenvalues of a Sturm-Liouville problem should be minima. As to the σ_g^2, since their inverse for the same reason as above are minima, one may expect them to be, asymptotically at least, maxima, this applying to their absolute values in the case of negative σ_g^2. One would then suspect σ_f^2 to be neither a true maximum or minimum. The inference seems to be supported by the results of Robe and Brandt (1966) who found, by applying the Rayleigh-Ritz method to the Chandrasekhar variational principle, that the exact σ_p^2 and σ_g^2 were approached respectively by decreasing and increasing values while in the case of σ_f^2, the successive approximations, at least in a fair proportion of cases, tend to the exact value both by larger and smaller values.

3. Non-Radial Adiabatic Oscillations of Various Models

3.1. THE HOMOGENEOUS INCOMPRESSIBLE SPHERE (Thomson, Lord Kelvin, 1863)

In that case, for any value of $l \geq 2$ ($l = 1$ has to be rejected here as it would correspond to a displacement of the centre of mass), there are $(2l+1)$ solenoidal modes with one and the same frequency

$$\sigma_l^2 = \frac{4\pi G\varrho}{3} l \frac{2(l-1)}{2l+1}.$$

In the nomenclature adopted above this is a f mode which is the only possible type in this model.

3.2. THE HETEROGENEOUS INCOMPRESSIBLE SPHERE

It is rather surprising that, although this is related to the simplest (no viscosity, no surface tension) spherical version of the Rayleigh-Taylor problem, detailed results have only been obtained recently. Perhaps the simplest case corresponds to an equilibrium configuration built of concentric layers of incompressible fluids of different densities. This has been considered recently by a student (Camps, 1973) who has accumulated quite a number of numerical results. As expected, apart from the f mode which is always present and is the continuation of the Kelvin mode of the homogeneous sphere, one finds as many g modes as there are discontinuities of densities. To each eigenvalue σ_g^2 corresponds an eigenfunction which in general reaches maximum amplitude (in absolute value) at the associated discontinuity (surface waves). The sign of σ_g^2 is the same as that of the density discontinuity $(\varrho_{in} - \varrho_{ex})$ with which it is associated, an instability, of course, developing at each interface where a heavier fluid (ϱ_{ex}) is superposed on a lighter one (ϱ_{in}). Thus, in general, when discontinuities of both signs are present, the g spectrum is split into two which we shall denote by σ_{g+}^2 and σ_{g-}^2 according as they correspond to stability or instability. In the case of four layers with two unstable discontinuities $(\varrho_{in} - \varrho_{ex} < 0)$ some peculiar behaviour was noted as the position and the value $(\varrho_{in} - \varrho_{ex})$ were varied at one of the unstable discontinuities. As the two e-folding times $(\propto 1/\sqrt{|\sigma_g^2-|})$ become close to each other, the eigenfunction normally associated with one discontinuity acquires a secondary extremum, sometimes quite important, at the position of the other discontinuity.

Another case was worked out by Robe (1974) as a by-product of his programme for polytropes since by letting the physical Γ_1 tend to infinity, one reduces the problem to that of the non-radial oscillations of a spherical configuration composed of an incompressible fluid with a continuously varying density. As $d\varrho/dr$ and $\mathscr{A}$ which reduces here to $(1/\varrho)(d\varrho/dr)$ are always negative in this case $(n>0)$, one gets a discrete infinite g spectrum with all positive eigenvalues σ_{g+}^2 in addition to the f mode. Another incompressible sphere with $\varrho = \varrho_c(1 + 0.1\, r^2/R^2)$ yielded a completely negative g spectrum as expected. Of course the p spectrum is suppressed (all $\sigma_p^2 \to \infty$) by the incompressibility.

3.3. THE HOMOGENEOUS COMPRESSIBLE SPHERE

This was the first compressible case treated in full details (Pekeris, 1938). The problem splits here into two second order differential equations, one for div $\delta\mathbf{r}$ and Poisson's equation which can be solved when the solution of the first equation is known. The latter can be expressed in polynomial form yielding an explicit biquadratic algebraic equation for the eigenvalues for a given l. There appear thus, in addition to the f mode which is identical here to the solenoidal Kelvin mode of the incompressible sphere with the same density (Robe, 1965), two families of modes, one with positive increasing eigenvalues as the number of nodes along r increases, the p modes, and one with negative eigenvalues tending to zero as the number of nodes along r increases, the g modes. That the latter are unstable (σ_g^2-) is not surprising since, in this case, $\mathscr{A}$ which reduces to $-(1/\Gamma_1 p)(dp/dr)$ is positive everywhere. As illustrated on Figure 1, the p modes go over smoothly into the radial modes ($l=0$) while, of course, the g and f modes have no counterparts for $l=0$.

3.4. THE POLYTROPES

The study of their non-radial oscillations was the occasion on which Cowling (1941) introduced the spectral classification used above. In their case

$$\mathscr{A} = \left(\frac{n}{n+1} - \frac{1}{\Gamma_1}\right) \frac{1}{p} \frac{dp}{dr},$$

where n is the polytropic index. If n and Γ_1 are constant throughout the star, $\mathscr{A}$ has a constant sign: positive (convective instability) if $(n+1)/n > \Gamma_1$, negative (convective stability) in the opposite case.

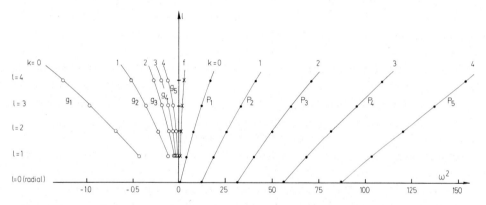

Fig. 1. Distribution of the discrete eigenvalues of the p (dots), f (crosses) and g (circles) modes for the homogeneous compressible model.

The advent of large electronic computers has made the numerical integration of even the 4th order problem a fairly simple matter especially for straight polytropes ($n=ct$) and many numerical results by Hurley et al. (1966) and by Robe (1968) have been added to the early results of Cowling (1941) and Kopal (1949). In particular, Robe (1968) has integrated the problem (both with and without Φ') for a whole series of polytropes ($n=0, 1, 2, 3, 3.25, 3.5, 3.75$ and 4) up to fairly high modes for various values of l and for $\Gamma_1 = 5/3$. Since $\Gamma_1 = 5/3$, only the polytropes $n=0$ and $n=1$ have $(n+1)/n > \Gamma_1$ and, as expected, they are the only one with a negative g spectrum. On the whole up to $n=3$, everything runs smoothly, the number of nodes which is always zero in δr, ϱ' and p' for the f mode goes to 1 for the same variables in the first p mode and increases regularly with the order of the mode to which it remains equal. The same is true of the g modes if they are stable. If they are unstable, the number of nodes in ϱ' and p' is still equal to the order of the mode but the number of nodes of δr is lower by one unit (cf. Table I).

However, some years ago, the work of Owen (1957) who had been unable to find modes without nodes for $n > 3.25$ raised the question of the existence of the f mode and of the first few p or g modes in models of high enough central condensation. However Robe's results show that all the modes continue to exist but they tend to acquire extra-nodes. For instance for $n = 3.25$, the f mode which has still zero node in δr and

TABLE I

Eigenvalues σ^2 in units of $\pi G\bar{\varrho}$, $l = 2$, $\Gamma_1 = 5/3$

Modes	$n=1$	Nodes of δr	Nodes of p'	Nodes of ϱ'	$n=3$	Nodes of δr	Nodes of p'	Nodes of ϱ'	$n=3.5$	Nodes of δr	Nodes of p'	Nodes of ϱ'	$n=4$	Nodes of δr	Nodes of p'	Nodes of ϱ'
p_{10}	416	10	10	10	312	10	10	10	297	10	10	10	288	10	10	10
p_9	346	9	9	9	262	9	9	9	252	9	9	9	245	9	9	9
p_8	282.2	8	8	8	217	8	8	8	210	8	8	8	206	8	8	8
p_7	224.5	7	7	7	176.5	7	7	7	171	7	7	7	171	7	7	7
p_6	173.2	6	6	6	139.8	6	6	6	137.2	6	6	6	140.3	6	6	6
p_5	128.2	5	5	5	107.4	5	5	5	106.8	5	5	5	116.6	5	5	5
p_4	89.44	4	4	4	79.23	4	4	4	80.39	4	4	4	102.2	4	6	6
p_3	57.11	3	3	3	55.29	3	3	3	57.88	3	3	3	83.83	5	5	5
p_2	31.32	2	2	2	35.63	2	2	2	39.68	2	2	2	67.75	4	4	4
p_1	12.41	1	1	1	20.35	1	1	1	27.91	1	1	1	56.18	3	5	5
f	1.997	0	0	0	10.90	0	0	0	21.55	2	2	2	45.77	4	4	4
g_1	−0.4039	0	1	1	6.553	1	1	1	16.13	1	3	1	36.79	3	5	5
g_2	−0.1844	1	2	2	3.771	2	2	2	11.39	2	2	2	30.67	4	4	4
g_3	−0.1070	2	3	3	2.430	3	3	3	7.655	3	3	3	23.99	5	5	5
g_4	−0.07029	3	4	4	1.694	4	4	4	5.417	4	4	4	20.48	4	4	6
g_5	−0.04989	4	5	5	1.248	5	5	5	4.023	5	5	5	17.02	5	5	5
g_6	−0.03731	5	6	6	0.958	6	6	6	3.099	6	6	6	13.44	6	6	6
g_7	−0.02899	6	7	7	0.759	7	7	7	2.464	7	7	7	10.78	7	7	7
g_8	−0.02319	7	8	8	0.616	8	8	8	2.00	8	8	8	8.82	8	8	8
g_9	−0.01898	8	9	9	0.510	9	9	9	1.66	9	9	9	7.34	9	9	9
g_{10}	−0.01583	9	10	10	0.429	10	10	10	1.40	10	10	10	6.20	10	10	10

ϱ' acquires 2 nodes in p'. For $n = 3.5$ not only does the f mode have 2 nodes in δr and ϱ' as well as in p' but the g_1 mode has acquired three nodes in p'. The situation deteriorates further as we go to higher central condensation and, for instance, for $n = 4$, the f mode has 4 nodes in all variables and the abnormality in the number of nodes extends to the p_4 and g_4 modes. One should also note that for high enough modes (p_5 and g_5 for $n = 4$) the number of nodes becomes regular again and equal to the order of the mode (cf. Table I).

The same phenomenon has now been found in physical models sufficiently evolved to have large central condensations (Dziembowski, 1971; Scuflaire, 1973). As stressed by Dziembowski and as we shall discuss later this can have very important consequences on other aspects of the problem like the question of vibrational stability.

The origin of this curious behaviour is however the same in physical models and polytropes for which it has been discussed by Robe (1968). This can be understood most easily on the basis of the approximate ($\Phi'=0$) second order system (9a-b) which, in the new variables

$$v = up^{1/\Gamma_1} \qquad w = y\varrho p^{-1/\Gamma_1}$$

becomes

$$\frac{dv}{dr} = \left[\frac{l(l+1)}{\sigma^2} - \frac{\varrho r^2}{\Gamma_1 p}\right] \frac{p^{2/\Gamma_1}}{\varrho} w \tag{12a}$$

$$\frac{dw}{dr} = \frac{1}{r^2}(\sigma^2 + \mathscr{A}g) \frac{\varrho}{p^{2/\Gamma_1}} v \tag{12b}$$

Let us first note that in general whatever the value of σ^2, both brackets in these equations will vanish at some point in the interval $0 < r < R$ since $\varrho r^2/\Gamma_1 p$ and $\mathscr{A}g$ vary from 0 at the centre to respectively $+\infty$ and $+$ or $-\infty$ (depending on the sign of $\mathscr{A}$) at the surface. This introduces some extra-singularities in the second order problem which tend to the surface or the centre as σ^2 becomes very large or very small. These singularities are however regular in Fuchs sense and, furthermore, they are apparent in the sense that all independent solutions and their derivatives remain finite and continuous across them. In particular, as the detailed discussion of the asymptotic behaviour of the eigensolutions shows (Vandakurow, 1967; Tassoul, 1967; Tassoul and Tassoul, 1968; Smeyers, 1968), the presence of these singularities does not invalidate our simplified arguments following Equations (9a-b) resting in particular on the neglect of terms in $1/\sigma^2$ (Equation (10)) or in σ^2 (Equation (11)) used to establish the characteristics of the p and g spectra. However, they may affect the behaviour of the solutions: whenever one of the involved bracket vanishes, the derivative of v (or w) vanishes too, just as it vanishes at the nodes of w (or v) and this can play havoc with the number and the distribution of zeros of consecutive modes at least for the first few modes in very condensed configurations. Indeed as the central condensation increases, $\mathscr{A}g$ develops a deeper and deeper minimum (cf. Figures 2 and 3) fairly close to the centre which implies that a greater and greater number of the low modes, starting with the f modes, have σ^2 values which make the factor $(\sigma^2 + \mathscr{A}g)$ vanish at three points in the star instead of one for low central condensation models. According to (12b), this forces w (or p') to change the sense of its variation and will, in favourable circumstances, bring about an extra zero. But this in turn, affects v (or δr) in a similar way and again an extra node may appear as discussed in detail for the polytrope $n=4$ in Robe (1968). However for high enough p modes (σ^2

NON-RADIAL OSCILLATIONS 145

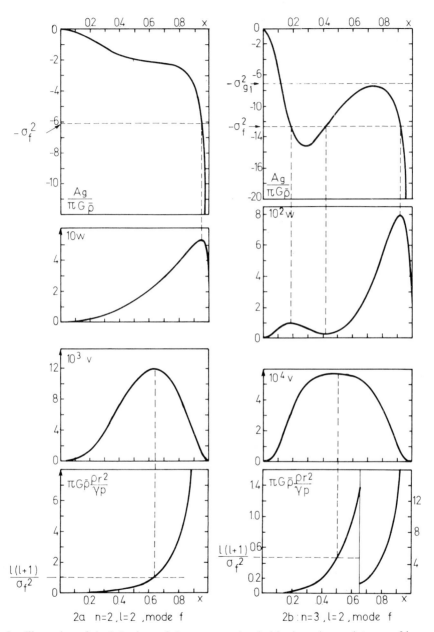

Fig. 2. Illustration of the behaviour of the apparent singularities, in various polytropes of increasing central condensation ($n=2$ and 3) and of the apparition of extra-nodes.

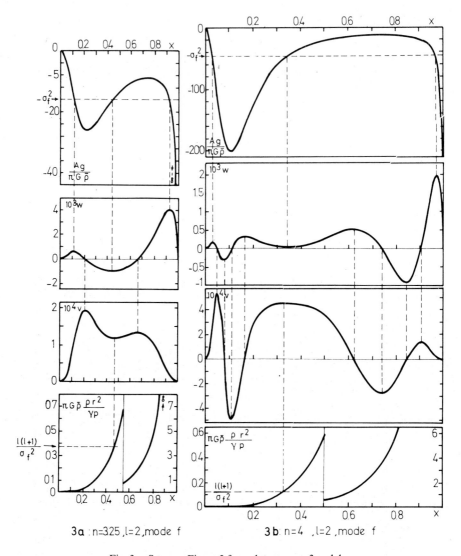

Fig. 3. Same as Figure 2 for polytropes $n=3$ and 4.

sufficiently large) or g modes (σ^2 sufficiently small) the situation always becomes normal again because, as in models with small central condensation, the factor $(\sigma^2 + \mathscr{A}g)$ finally vanishes again at one point only.

The minimum of $\mathscr{A}g$ tends to become deeper and to approach the centre as the central condensation increases as illustrated on Figure 4 where the behaviour of $\varrho r^2 / \Gamma_1 p$ is also represented as well as the positions of a few apparent singularities for the polytrope $n=4.2$ and a fairly highly evolved model.

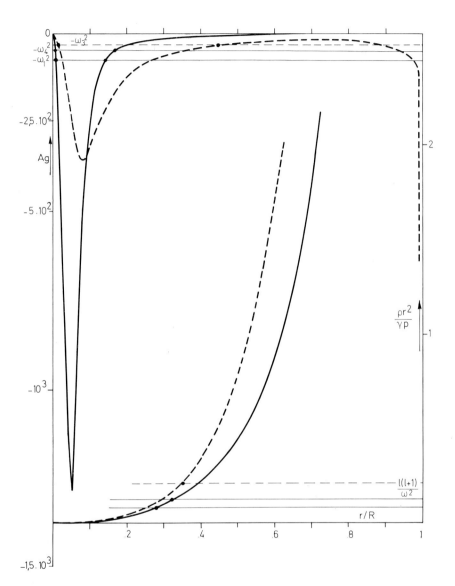

Fig. 4. Illustration of the behaviour of $\mathscr{A}g$ and $\varrho r^2/\Gamma_1 p$ in highly condensed configurations with the positions of a few apparent singularities. Polytrope $n=4.2$: broken lines; evolved model: full lines. In the latter $\mathscr{A}g$ tends to remain close to zero up to very near the surface because of the extensive external convection zone (after Scuflaire, unpublished).

The behaviour of p and g modes, taken from Robe's study, is illustrated on Figure 5 for various polytropes and $l=2$. The p modes behave very much like the radial modes with the ratio of the amplitudes at the surface to that near the centre increasing very much as the central condensation increases. On the other hand, as l increases, the

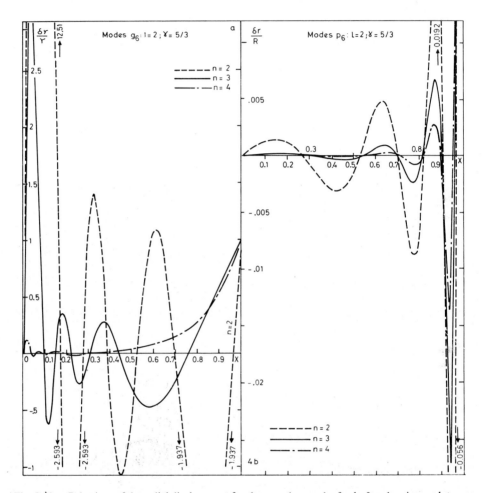

Fig. 5a–b. Behaviour of the radial displacement for the g_6 and p_6 modes for $l=2$ and various polytropes.

nodes of the p modes tend to concentrate in the external layers as already noted by Smeyers (1967). This is illustrated on Figure 6 for the p_3 mode of the polytrope $n=3$ according to Robe (1973b). The g modes offer more peculiar characteristics, the amplitude being susceptible to reach appreciably higher values in the interior than at the surface at least as long as the central condensation is not too large. On the other hand, in the case of very high central condensation, the occurrence of extra-nodes

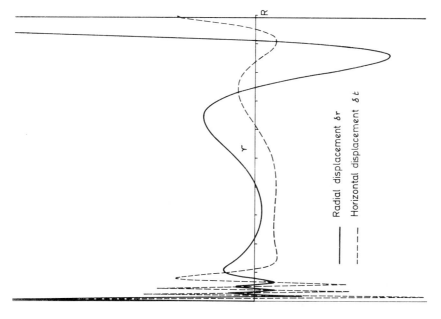

Fig. 7. Run of the amplitude for the radial and horizontal displacements in the case of the g_4 mode of a highly condensed physical model ($\varrho_c/\bar{\varrho} \simeq 3 \times 10^3$) (after Scuflaire, unpublished).

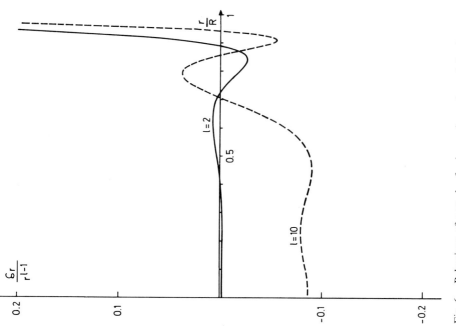

Fig. 6. Behaviour of p modes for increasing values of the degree l of the spherical harmonic.

close to the centre seems to favour also a pick-up of the amplitude in these regions (cf. Figure 7 from Scuflaire's work).

3.5. Physical models

The problem has been integrated numerically for main-sequence models of fairly high mass by Van der Borght and Wan Fook Sun (1965) and by Smeyers (1967). Since these models of fairly low central condensation possess very extensive convective cores where $\mathscr{A}$ is essentially equal to zero, it is natural to find, apart from the f and

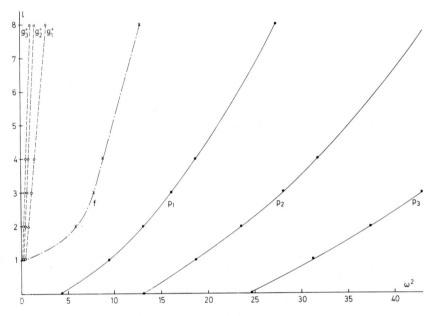

Fig. 8. Distribution of the eigenvalues of the p (dots), f (crosses) and g (circles) modes for a main sequence model of fairly high mass.

p modes, which behave very much as in polytropes of index $n=3$, families of g^+ modes with eigenvalues concentrated in a very small interval around zero (cf. Figure 8) since if $\mathscr{A}$ was zero everywhere all the σ_g^2 would vanish. The behaviour of these g modes is also peculiar, the amplitude oscillating only in the radiative envelope ($\mathscr{A}>0$) while it decreases more or less exponentially in the convective core (cf. Figure 9).

As already mentioned, the non-radial oscillations of quite a variety of stellar models at different stages of evolution have been studied in recent years (Dziembowski, 1971; Robe *et al.*, 1972; Dziembowski and Sienkiewicz, 1973; Scuflaire, 1973; Osaki, 1973) mostly in connection with various aspects of vibrational stability towards such perturbations but examples of various adiabatic modes will be found there.

Table II summarizes in general terms the various types of possible spectra for the various models considered here, illustrating their origin and their lineage.

For those interested in the possible applications of non-radial oscillations to the interpretation of periodic variability in certain stars, Tables III and IV present re-

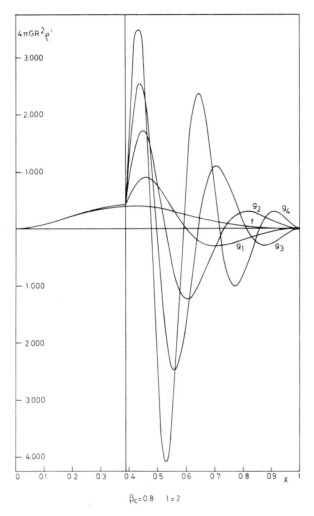

Fig. 9. Run of ϱ' for various g modes in a main sequence model of fairly high mass with a large convective core.

spectively values of $Q = P_{\text{days}} (\bar{\varrho}/\bar{\varrho}_\odot)^{1/2}$ for massive stars (Smeyers, 1967, cf. also Figure 8) for different values of l and for various polytropes (Robe, 1968) for $l=2$. In some cases, as we shall see later, the jump in period from the f and p modes to the g modes may be much larger (cf. Table VI).

TABLE II

Types of spectra for various models

(X: exists; (X): exists but has to be rejected; O: does not exist)

Models	Incompressible				Compressible			
	Homogeneous		Heterogeneous		Homogeneous		Non-homogeneous	
	$l=1$	$l>1$	$l=1$	$l>1$	$l=1$	$l>1$	$l=1$	$l>1$
p modes non-solenoidal infinite discrete for l fixed	O	O	O	O	X	X	X	X
f mode solenoidal non-solenoidal, one mode only for l and m fixed	(X) O	X O	(X) O	X O	(X) O	X O	(X) O	O X
g modes solenoidal non-solenoidal	O O	O O	(X) O discrete finite *or* infinite	X O	O X	O X infinite	O X discrete	O X

TABLE III

Q values. $(M/M_\odot)\bar{\mu}^2 = 9.85$, $\beta_c = 0.8$, $\varrho_c/\bar{\varrho} = 20.07$

Degree sph. harm. Modes	$l=0$ (radial)	$l=1$	$l=2$	$l=3$	$l=4$	$l=8$
g_4	–	0.6630	0.3882	0.2794	0.2208	0.1288
g_3	–	0.5244	0.3093	0.2245	0.1791	0.1088
g_2	–	0.3829	0.2292	0.1693	0.1376	0.0898
g_1	–	0.2378	0.1477	0.1137	0.0963	0.0718
f	–		0.0514	0.0427	0.0392	0.0323
p_1	0.0566 (fund.)	0.0377	0.0322	0.0289	0.0268	0.0223
p_2	0.0318 (1st mode)	0.0268	0.0238	0.0219	0.0206	0.0174
p_3	0.0233 (2nd mode)	0.0208	0.0190	0.0177	0.0168	0.0145
p_4		0.0170	0.0158	0.0149	0.0142	0.0124

TABLE IV

Q values for polytropes ($\Gamma_1 = 5/3$, $l=0$: radial and $l=2$)

$n=2$ $\varrho_c/\bar{\varrho}=11.40$		$n=3$ $\varrho_c/\bar{\varrho}=54.18$		$n=3.5$ $\varrho_c/\bar{\varrho}=152.9$		$n=4$ $\varrho_c/\bar{\varrho}=622.4$	
Radial	Non-radial	Radial	Non-radial	Radial	Non-radial	Radial	Non-radial
	g_{10} : 0.664		g_{10} : 0.204		g_{10} : 0.113		g_{10} : 0.0536
	g_5 : 0.384		g_5 : 0.120		g_5 : 0.0668		g_5 : 0.0325
	g_4 : 0.327		g_4 : 0.103		g_4 : 0.0575		g_4 : 0.0296
	g_3 : 0.270		g_3 : 0.0851		g_3 : 0.0484		g_3 : 0.0273
	g_2 : 0.213		g_2 : 0.0681		g_2 : 0.0397		g_2 : 0.0242
	g_1 : 0.154		g_1 : 0.0523		g_1 : 0.0333		g_1 : 0.0221
	f : 0.06572		f : 0.04056		f : 0.02884		f : 0.01979
fund: 0.0563	p_1 : 0.0341	fund: 0.0381	p_1 : 0.0297	fund: 0.0326	p_1 : 0.0253	fund: 0.0298	p_1 : 0.0179
(1): 0.0317	p_2 : 0.0236	(1): 0.0281	p_2 : 0.0224	(1): 0.0252	p_2 : 0.0213	(1): 0.0232	p_2 : 0.0163
(2): 0.0226	p_3 : 0.0182	(2): 0.0217	p_3 : 0.0180	(2): 0.0205	p_3 : 0.0178	(2): 0.0190	p_3 : 0.0146
(3): 0.0175	p_4 : 0.0148	(3): 0.0176	p_4 : 0.0150	(3): 0.0171	p_4 : 0.0149	(3): 0.0162	p_4 : 0.0132
(4): 0.0143	p_5 : 0.0125	(4): 0.0148	p_5 : 0.0129	(4): 0.0147	p_5 : 0.0130	(4): 0.0140	p_5 : 0.0124

3.6. Models with two or more regions of opposite signs in $\mathscr{A}$

We have already recalled Lebovitz's result (1965, 1966) according to which the presence in a star of any finite region, however small, with $\mathscr{A}>0$ implies the existence of negative g eigenvalues $(\sigma_{g^-}^2)$. But it would be surprising that, in such cases, positive $\sigma_{g^+}^2$ would not also subsist. Indeed it is easy to verify (Ledoux and Smeyers, 1966) that Equation (11) can be written in self-adjoint form as

$$\frac{d}{dr}\left(\varrho\frac{du}{dr}\right)+u(\lambda s-t)=0 \tag{13}$$

with

$$\lambda=-\frac{1}{\sigma^2}, \quad s=\frac{l(l+1)\,\mathscr{A}g\varrho}{r^2}, \quad t=\frac{l(l+1)\varrho}{r^2}-\varrho\frac{d}{dr}\left(\frac{1}{\Gamma_1 p}\frac{dp}{dr}\right)$$

the solution u in which we are interested and its derivative having to be regular in the complete interval $0 \leqslant r \leqslant R$. But it is well known (cf. Ince, 1964) that, if in this equation s changes sign in the interval $0<r<R$, there are indeed two spectra. If t is everywhere positive, which is generally true in stellar models, then one spectrum is entirely positive while the other is entirely negative and in both cases $|\lambda|$ tends to infinity by discrete values. Converting to $\sigma^2\,(=-1/\lambda)$, we keep of course two spectra, entirely negative or positive, but converging by discrete values to zero.

As far as the eigensolutions are concerned, Equation (13) shows that the g^+ modes (stable, $\lambda<0$, $\sigma^2>0$) do not oscillate in the unstable region ($\mathscr{A}>0$, $s>0$) where the amplitude tends to decrease exponentially while all the nodes tend to accumulate in the stable region ($\mathscr{A}<0$, $s<0$). The reverse is true of the g^- modes (unstable, $\lambda>0$, $\sigma^2<0$) which do not oscillate in the stable region ($\mathscr{A}<0$, $s<0$) (cf. Figure 10).

This behaviour was checked numerically by Smeyers (1966) for the complete 4th order problem on artificial models, for instance with a strongly convectively unstable core ($\mathscr{A}>0$) and a strongly convectively stable envelope ($\mathscr{A}<0$).

The analysis of asymptotic non-radial modes by Tassoul and Tassoul (1968) which is certainly adequate in the case of only two regions of opposite signs for $\mathscr{A}$ (only one turning point at the junction) gives another very clear illustration of the existence and the behaviour of these two g spectra. Let us recall also that even the first g^+ modes in the large mass stars of Smeyers with massive cores in convective equilibrium ($\mathscr{A}\simeq 0$) had also a similar behaviour. In this case, using the small but positive values of $\mathscr{A}$ in the core, necessary to insure the heat transfer, it was also possible to isolate the negative g spectrum corresponding to unstable g^- modes oscillating only in the core and decaying monotonically and rapidly in the external radiative envelope. Such g^- modes have also been studied by Saslaw and Schwarzschild (1965) (cf. also Fowley, 1972) to discuss the penetration of convection from the core into the surrounding stable envelope. Even if this penetration is very small, there are still interesting questions related to the possible excitation by these unstable g^- modes of other modes (g^+ and p modes) and the possible transport of heat by these modes in the

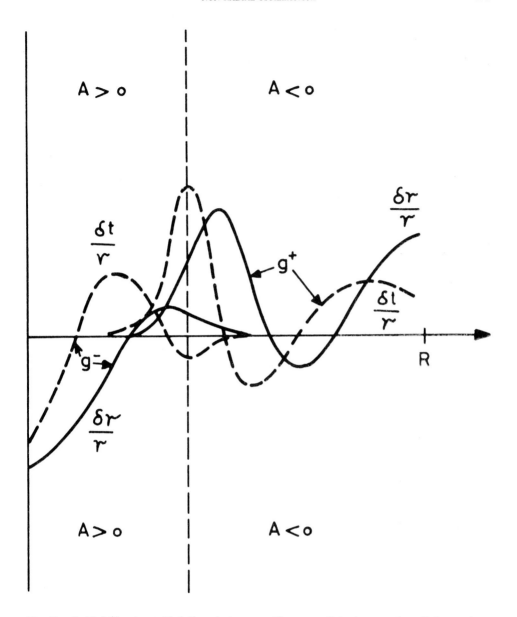

Fig. 10. Stable (g^+) and unstable (g^-) modes in a star with a super adiabatic core and a radiative envelope.

transition zone in the stable region across which they get damped. A particularly interesting problem would arise if some of these other modes turned out to be vibrationally unstable locally (cf. for instance Spiegel (1964), for such a vibrational instability of p modes in a superadiabatic region, also Moore and Spiegel (1966); for g^+ modes, Souffrin and Spiegel (1967)) so that they could pick up extra energy in that region enabling them to move further out before being damped, achieving a more efficient heat transfer. A similar idea has been discussed by Ulrich (1970) in another context to explain the heating of the solar chromosphere and the corona by dissipation, in these external layers, of the 5-min oscillation considered as a standing acoustic wave trapped at subphotospheric levels.

On the other hand, it is well known (Cowling and Newing, 1949; Ledoux, 1949, 1951) that in presence of a slow rotation Ω, the $(2l+1)$ degeneracy of the eigenvalues is lifted for all the modes so that the eigenvalues in a rotating frame become

$$\sigma_{l,m} = \sigma_l + mC_l\Omega \quad -l \leqslant m \leqslant l,$$

where σ_l is the square root of the eigenvalue (σ_l^2) of the non-rotating star for the same value of the degree l of the spherical harmonic and C_l, a constant which can be evaluated from the corresponding eigensolution. If the σ_l considered is that of a g^- mode $(\sigma_l^2 < 0)$ it is pure imaginary and the resulting motion tends to be oscillatory with a period proportional to that of rotation and an increasing amplitude fed by the available buoyancy energy. If the convection zone in which these g^- modes are localized is in the external layers, Ledoux (1967) has suggested that an oscillatory theory of magnetic variable stars could be developed on this idea.

Of course, resonance between one of these overstable convective mode and another dynamically stable g^+ mode for instance could help stabilize both the period and the motion. This last possibility has been suggested recently in a very interesting paper by Osaki (1974) in an effort to find an explanation to the origin of the variability of the β-Cephei stars for which up to now all other sources of pulsational instability for radial as well as non-radial modes have failed. However, in this case, the unstable g^- modes considered by Osaki are localized in the important and fairly rapidly spinning convective core of models in the late stages of the main sequence which is not unreasonable for β-Cephei stars.

He shows that a resonance between the overstable convective mode with a period determined by that of the rotation of the core and the non-radial f mode for $l=2$, $m=2$ is possible. On the other hand, he finds that, due to the already fairly high central condensation of the model, this f mode has one node somewhat outside the convective core and fairly large amplitudes in the core. Of course, this is favourable since it is the coupling, which occurs linearly in presence of rotation, with the g^- mode localized in the core which is supposed to drive the f mode. This would lead to a preferentially excited wave traveling around the equator in the same direction as the rotation as suggested initially by Ledoux (1951), a hypothesis still found to be the most plausible in a previous paper of Osaki (1971) as far as the beat phenomenon and the characteristic variations in line width of these stars are concerned. The Q value, 0.033, given by Osaki for the f mode seems reasonable for a somewhat condensed model (cf. Table IV, n_{eff} between $n=3$ and $n=4$) but it may still be somewhat high since direct evaluations from observations seem to lean towards smaller values (0.027 in Ledoux and Walraven, 1958; 0.021 in Hitotuyanagi and Takeuti, 1964). Nevertheless, it does not seem excluded that the same mechanism could work with one of the first p mode which have lower Q values.

If the convective zone is in the very external layers of small density and small heat capacity, the subsisting superadiabatic gradient, once turbulent convection is established at a dominant relatively small wavelength, may be much larger than in the central core. In such cases, one might expect erratic excitation of g^- modes of long wavelengths giving rise to transient phenomena observable at the surface. As above, in presence of rotation or magnetic field providing a restoring force, this may also give rise to transient or semi-permanent oscillatory motions with periods determined by the rotation or the magnetic field.

In the sun, excitation, by these g^- modes in the hydrogen convection zone, of acoustic p waves which get dissipated as shock-waves in the tenuous layers at higher altitude is often advocated as the heating mechanism of the chromosphere and corona but we cannot go here into the many and complex aspects of these theories. Another interesting problem for the Sun atmosphere has been raised by the discovery by Leighton (1961) of a 5-min periodicity of the gas velocity in the photosphere and the chromosphere. It seems, however, that the complexity of the external layers of the Sun has allowed many different suggestions, all with a certain degree of plausibility and some with special advantages (cf. the review by Schatzman and Souffrin, 1967). The last four known to the present author ranges from trapped acoustic waves well below the photosphere (Ulrich, 1970) or even straight standing p modes (Wolff, 1972) to trapped gravity waves (g^+ modes) between the top of the hydrogen convection zone and the hydrogen ionization zone in the chromosphere (Thomas *et al.*, 1971) through trapped acoustic waves in the chromosphere (McKenzie, 1971) all four recovering at least the 5-min periodicity. A careful numerical analysis of all the possible modes in as good a model as possible of a fairly extensive outer envelope of the Sun perhaps including a good part or the whole convection zone of hydrogen and taking into account the variation of the mean molecular weight $\bar{\mu}$ as well as the new ionization of hydrogen in the upper chromosphere would be very welcome.

Cases of more than two regions of alternating signs in $\mathscr{A}$ are also of interest as such situations arise in many models in the course of stellar evolution. The heterogeneous incompressible model composed of superposed layers of different densities discussed previously offers the simplest example and we already noted there that, in the case of two unstable discontinuities $(\varrho_{\text{in}} - \varrho_{\text{ex}}) < 0$ separated by a stable one, the behaviour of the eigenfunctions associated with the two corresponding negative eigenvalues $\sigma_{g^-}^2$ can drastically depend on the closeness of these eigenvalues. While, in general, i.e. as long as they are not very close, each of the eigenfunctions has a single maximum at the unstable interface with which it is associated, one of them may acquire a secondary important maximum at the other discontinuity, the minimum amplitude between the two remaining appreciable, when its eigenvalue becomes very close to that associated with the other discontinuity.

Goosens and Smeyers (1974), on the other hand, have studied a compressible composite polytropic model consisting of two convectively stable zones separated by a convectively unstable one. Apart from the unstable g spectrum (σ_{g-}^2) associated with the intermediate zone, they also find two stable g spectra, one essentially associated with the core, the other with the envelope. Again, in general, a mode associated with one given region has an appreciable amplitude in that region only and stable modes do not oscillate in the intermediate unstable region while unstable modes do not oscillate in the surrounding stable regions. However, for the stable modes, here again they find more or less accidental 'resonances' giving rise to stable modes with an appreciable amplitude in the two stable regions which contribute nearly equally to the eigenvalues of these modes.

This last phenomenon does not appear in Tassoul and Tassoul (1968) asymptotic discussion which associates with each region of a given sign for $\mathscr{A}$ a spectrum of g modes (stable or unstable depending on the sign of $\mathscr{A}$) with amplitude large in that region only and decreasing exponentially outside. But if we look back at Equation (13) we see that the points where $\mathscr{A}$ vanishes are turning points of the equation in Langer's terminology, i.e. points where the coefficient (s) of the eigenvalue vanishes. Now the method used by Tassoul and Tassoul which is valid when there is only one turning point (2 regions of opposite signs for $\mathscr{A}$) may lead to partly erroneous or incomplete results in the case of two turning points (three regions of alternating signs for $\mathscr{A}$) according to Langer (1959) who, in this case, develops the solution in terms of Weber functions rather than Bessel functions as in the case of one turning point. It may be that this is the reason why the more or less resonant interaction between two zones does not appear in the discussion of Tassoul and Tassoul. Indeed, a preliminary investigation by Iweins and Ledoux (1971) although not using Weber functions but treating the various junctions more completely than in Tassoul and Tassoul, already suggested something of the sort when the eigenvalues of the two regions were commensurable.

I believe that this is an important question which should be pursued because of its possible applications to the case encountered in stellar evolution when two convective zones approach each other sometimes very closely since the penetration through the narrow intermediate stable zone may be sufficiently enhanced to lead really to a complete remixing of the whole layer which could have important effects on the evolution.

4. Vibrational Stability Towards Non-Radial Modes

As we mentioned at the beginning, it is only relatively recently that active interest in this question arose. As is well known, the problem here is to evaluate the effects of the non-conservative terms in Equations (1b) and (1c) which have been neglected up to now. They are usually very small compared to the other terms in the equations except perhaps in a fairly narrow external region whose heat capacity is often sufficiently small to render its influence negligible. In a first order theory, they introduce a damping factor in the time dependence which becomes $e^{i\sigma_a t} \cdot e^{-\sigma' t}$, the perturbation being vibrationally stable or unstable depending on whether σ' is positive or negative.

As long as the σ_a^2 of the adiabatic mode considered does not become too small (which may happen for the highest g modes), one may, as in the case of radial modes, use a perturbation method to evaluate σ'. As shown by Simon (1957) when the unperturbed model is in hydrostatic and thermal equilibrium, this yields

$$\sigma' = -\frac{1}{2\sigma_a^2} \frac{\iiint (\Gamma_3-1)\left(\frac{\delta\varrho}{\varrho}\right)_a \left(\varepsilon_N - \varepsilon_\nu - \frac{1}{\varrho}\operatorname{div} \mathbf{F}\right)_a' \varrho r^2 \sin\theta \, dr \, d\theta \, d\varphi}{\iiint (\boldsymbol{\delta r} \cdot \boldsymbol{\delta r}^*)_a \varrho r^2 \sin\theta \, dr \, d\theta \, d\varphi}, \qquad (14)$$

where the suffix 'a' denotes solutions of the adiabatic problem or terms to be evaluated by means of these solutions.

In this expression, we have written explicitly $\varepsilon = \varepsilon_N - \varepsilon_\nu$ where ε_N is the rate at which nuclear energy is transformed into heat and ε_ν represents the net rate of neutrinos emission at the expense of the thermal energy. However, we have still neglected the effects of viscosity which is certainly justified as far as molecular and radiative viscosity are concerned but may not be the case at all if turbulent viscosity comes into play (Counson et al., 1956). On the other hand, in the latter case, the transfer of momentum would also probably be sufficiently important to modify appreciably the distribution of the amplitudes in the star and this leads to a much more difficult problem.

Let us assume that $\mathbf{F}$ reduces to the radiative flux

$$\mathbf{F}_R = -\frac{c}{\kappa\varrho}\operatorname{grad}\left(\tfrac{1}{3}aT^4\right)$$

and let us denote the sensitivity of ε_N and κ to ϱ and T by the coefficients

$$(\varepsilon_N)_\varrho \equiv \varepsilon_\varrho = \left(\frac{\partial \log \varepsilon_N}{\partial \log \varrho}\right)_T, \quad \varepsilon_T = \left(\frac{\partial \log \varepsilon_N}{\partial \log T}\right)_\varrho, \quad \kappa_\varrho = \left(\frac{\partial \log \kappa}{\partial \log \varrho}\right)_T, \quad \kappa_T = \left(\frac{\partial \log \kappa}{\partial \log T}\right)_\varrho$$

the neutrino losses being neglected in the following although they can easily be reintroduced by analogy with ε_N.

As all perturbations in (14) depend on θ and φ through known functions $P_l^m(\cos\theta) \times e^{im\varphi}$ and their derivatives, the integration on these variables may be carried out yielding, for a mode corresponding to a spherical harmonic of degree l

$$\sigma'_l(2\sigma_{l,a}^2 I) = -\int_0^R \left[\left(\frac{\delta T}{T}\right)^2_{l,a}\left(\frac{\varepsilon_\varrho}{\Gamma_3 - 1} + \varepsilon_T\right) + \left(\frac{\delta T}{T}\right)_{l,a}\frac{l(l+1)}{\sigma_{l,a}^2 r^2}\chi_{l,a}\right] 4\pi\varrho\varepsilon r^2 \, dr +$$

$$+ \int_0^R \left(\frac{\delta T}{T}\right)_{l,a}\frac{d}{dr}\left\{L(r)\left[4\left(\frac{\delta r}{r}\right)_{l,a} + (4-\kappa_T)\left(\frac{\delta T}{T}\right)_{l,a} - \kappa_\varrho\left(\frac{\delta\varrho}{\varrho}\right)_{l,a} + \right.\right.$$

$$+ \frac{(d/dr)(\delta T/T)_{l,a}}{(1/T)(dT/dr)} - \frac{l(l+1)\chi_{l,a}}{\sigma^2 r^2}\Bigg]\Bigg\} dr +$$

$$+ \int_0^R \left(\frac{\delta T}{T}\right)_{l,a} (\delta r)_{l,a} \frac{l(l+1) 16\pi ac T^4}{3\kappa\varrho} \left(\frac{\Gamma_2-1}{\Gamma_2} - \frac{d\log T}{d\log p}\right) \frac{1}{p} \frac{dp}{dr} dr +$$

$$+ \int_0^R \left(\frac{\delta T}{T}\right)_{l,a} \left(\frac{p'}{p}\right)_{l,a} \frac{\Gamma_2-1}{\Gamma_2} l(l+1) \frac{16\pi ac T^4}{3\kappa\varrho} dr, \tag{15}$$

where

$$I = \int_0^R \left(\delta r_{l,a}^2 + \frac{l(l+1)}{\sigma_a^4 r^2} \chi_{l,a}^2\right) 4\pi\varrho r^2 \, dr$$

and

$$\chi = \left(\frac{p'}{\varrho} + \Phi'\right), \quad L(r) = 4\pi r^2 F_r(r),$$

where Φ' is often negligible.

This expression recalls the familiar form of the damping coefficient for radial pulsations, the essential differences coming from the terms in $l(l+1)$ which correspond to the effects of the inequalities of ϱ' and T' on a level surface introduced by the factor $P_l^m(\cos\theta) e^{im\varphi}$ and the resulting horizontal gradients.

The first term corresponds to the effects of the energy generation and, if it is concentrated around the centre as is often the case, its importance, as already pointed out by Simon (1957), is less than in the radial case as all non-radial amplitudes δT, $\delta\varrho$, δp tend to zero at the centre while they remain finite in radial oscillations. One may then expect that the vibrational stability of ordinary stellar models and especially of main sequence models towards non-radial perturbations will be easily secured by the generally stabilizing influence of the radiative conductivity represented by the second term, as confirmed by Wan's discussion (1966).

The third term is really characteristic of non-radial perturbations. In particular, it is proportional to the argument $\mathscr{S}$ (cf. 2′) of Schwarzschild's criterion according to which, in case of uniform chemical composition, radiative equilibrium is stable if $\mathscr{S} < 0$. In this case, this term favours vibrational stability since $(\delta T/T)_a$ and $(\delta r)_a$ being generally of opposite signs while dp/dr is negative, its contribution to σ' is positive. However, in a superadiabatic region the sign of $\mathscr{S}$ is inversed and this term may contribute negatively to σ' favouring vibrational instability. This must correspond to some of the effects discussed, as already mentioned, by Spiegel (1964) and by Souffrin and Spiegel (1967).

On the other hand, in presence of a gradient of chemical composition, the mean molecular weight $\bar{\mu}$ decreasing towards the exterior, stable radiative equilibrium subsists as long as $\mathscr{A} < 0$ even if $\mathscr{S} > 0$ (cf. 2"). Thus such a zone where

$$\frac{\Gamma_2 - 1}{\Gamma_2} \frac{d \log T}{d \log p} < \frac{\Gamma_2 - 1}{\Gamma_2} + \frac{\beta}{4 - 3\beta} \frac{d \log \bar{\mu}}{d \log p} \qquad (16)$$

should contribute to vibrational instability as shown by Kato (1966) using a local discussion.

As far as the last term is concerned, it is negligible for high g modes (p', small) and it remains probably always rather small, contributing mainly to vibrational stability, since more often than not p' and δT are of the same sign.

When radial modes are unstable as in Cepheids and RR Lyrae stars due to the special behaviour of κ and Γ in the ionization zones of H and He in the very external layers, one might expect (cf. Zahn, 1968) that non-radial p modes should also be vibrationally unstable since the run of the amplitude is generally similar for the two types of modes. However, as pointed out by Dziembowski (1971) the central condensation of the appropriate models is already very high and causes, as we have recalled previously, the apparition in the p eigenfunctions of extra-nodes in the central region with an increase of the amplitude which thus varies very rapidly, enhancing quite strongly the conductive dissipation in that region. According to Dziembowski the last effect is strong enough to damp the distabilizing influence of the external layers. However this may not be true of all p-modes especially those of large l as discussed above (cf. Figure 6).

On the other hand, as we have stressed before, g^+ modes can have fairly large amplitudes in the interior favouring for instance the distabilizing effects of nuclear reactions especially if the latter are not concentrated at the centre. In particular, according to their behaviour in models comprising two regions of opposite signs in $\mathscr{A}$, g^+ modes should have large amplitudes in the small radiative core of a small mass star decreasing rapidly in its convective envelope so that the effect of the core where the energy generation takes place should be dominant. The vibrational stability of such a model for $M = 0.5\ M_\odot$ in its early main sequence phase was studied by Robe et al. (1972). Although, as expected, the vibrational stability of the g^+ modes, especially g_1^+ for $l=2$, decreased with the radius of the core, it still subsisted for the smallest core radius reached ($r/R = 0.354$) although the positive contribution to σ' of term (2) in (15) exceeded only very slightly the absolute value of the negative contribution of term (1). In fact, a somewhat higher sensitivity of ε to T or ϱ or a lower sensitivity of κ could easily have brought about vibrational instability. In fact Noels et al. (1974) find that the effective ε_T is appreciably larger than the one used by Robe et al. and sufficient to lead to instability. One should note however that as the horizontal component of the g^+ modes considered is also important, the corresponding unstable motion could at most lead, when sufficiently amplified, to some kind of forced turbulent semi-convection. In the case of an inhomogeneous core, the corresponding mixing could be significant because of the induced turbulence and might have important evolutionary consequences.

One might think also that late evolutionary phases, when the models acquire large convective envelopes again and develop shell-source of energy at some distance from the centre, would offer even better conditions for vibrational instability of these g^+ modes. An investigation of this problem has been started in Liège, but it ran into the already discussed difficulty of the appearance of a great many extra-nodes in the central region due to the very high central condensation which, as pointed out by Dziembowski (1971) will increase considerably the dissipation there. Nevertheless, until computations are more advanced it is impossible to come to a definite conclusion in this case.

5. The Solar Neutrino Problem and Local Analysis

The question of mixing as a result of the vibrational instability of some g^+ modes has been raised by Dilke and Gough (1972) in an attempt to find a solution to the solar neutrino problem. Following a suggestion by Fowler (1972), different authors (Rood, 1972; Ezer and Cameron, 1972) have shown that if, after the Sun has evolved sufficiently to possess an appreciable positive gradient of H and especially ^{3}He, the central half or so of it is suddenly mixed, the following readjustments lead finally to an expansion of the central part and a lowering of the neutrino flux to something like the present observed upper limit, a situation which should not last very much longer than a few million years while the luminosity is around its minimum. While the cause of the mixing was left undetermined in these two investigations, Dilke and Gough on the contrary attempted to relate it to the vibrational instability of a g^+ mode.

Of course, this instability must originate here in the gradient of chemical composition set up by the evolution and more particularly in the gradients of H and especially ^{3}He. This would have the great advantage that, the critical values of these gradients necessary for initiating mixing being fixed by the vibrational instability criterion, it would become possible to estimate the time elapsing between successive mixings, i.e. the time necessary to build these critical gradients, and perhaps compare this to the time separating major glaciations since the sun luminosity would also drop to a minimum after each mixing.

The qualitative argument of Dilke and Gough in favour of this g^+ mode vibrational instability was based on a local discussion. This type of approach has been used a few times for different purposes in recent years and since it may at least point to interesting factors affecting the stability, we may as well summarize the results. In such approaches, the general equations are usually particularized to a narrow layer of negligible curvature $(r \to \infty)$, $l(l+1)/r^2 \to k_H^2$, where k_H is the horizontal wave-number in the plane (x, y). One may then to a very good approximation neglect Φ' and, for fairly high g modes, p' as well except in the equation of motion. Usually the fluid is treated as incompressible

$$\text{div}\,\delta\mathbf{r} = 0 \tag{17}$$

except for the effects of thermal dilatation (Boussinesq approximation). Some simplifying assumption is often made also concerning the conductivity, but here we shall let K in the heat equation

$$\mathbf{F} = -K \operatorname{grad} T$$

vary quite generally with ϱ and T.

So as to cover the case treated by Dilke and Gough (pp-chain), we shall write

$$\varepsilon' = \varepsilon \left(v \frac{T'}{T} + \frac{\varrho'}{\varrho} \right) + \mu_1 \frac{X'_1}{X_1} + \mu_3 \frac{X'_3}{X_3},$$

where v is the usual effective exponent characterizing the sensitivity of the nuclear reactions to changes in T on the time-scale of the perturbation (our previous ε_T; for the pp-chain $\varepsilon_\varrho = 1$); X_1 and X_3 denote the abundances in mass of H and ^{3}He entering the reactions $p(p, \beta^+ v) D(p, \gamma)$ ^{3}He and ^{3}He(^{3}He, 2p) ^{4}He retained as the most significant by these authors and releasing energy at the rate ε_{11} and ε_{33} respectively. We then have

$$v = (v_{11}\varepsilon_{11} + v_{33}\varepsilon_{33})/\varepsilon, \quad \mu_1 = 2\varepsilon_{11}/\varepsilon, \quad \mu_3 = 2\varepsilon_{33}/\varepsilon$$

and the energy Equation (1c''') may be written

$$T' - T\delta r \mathcal{S} = \frac{1}{sC_p} \left(\varepsilon + \frac{1}{\varrho} \operatorname{div}(K \operatorname{grad} T) \right) \tag{18}$$

where $i\sigma$ has been replaced by s since we shall adopt here a time dependence of the form e^{st}.

As the variation of the chemical composition from point to point is an important factor in the problem, we must add explicitly the following three equations

$$\bar{\mu}' + \delta z \frac{d\bar{\mu}}{dz} = 0, \quad X'_1 + \delta z \frac{dX_1}{dz} = 0, \quad X'_3 + \delta z \frac{dX_3}{dz} = 0 \tag{19}$$

which express the fact that the chemical composition does not vary following the motion which is too fast for transmutations to be significant. Finally, the equation of state (gas + radiation) taking into account the variations of chemical composition yields:

$$\frac{\beta \varrho'}{\varrho} - \beta \frac{\bar{\mu}'}{\bar{\mu}} + (4 - 3\beta) \frac{T'}{T} = 0, \tag{20}$$

where $\beta = p_G/p$ and μ has its usual definition

$$\bar{\mu} = \frac{2}{1 + 3X_1 + 0.5X_4}$$

from which $\bar{\mu}'$ may be computed in terms of X'_1. Together, Equations (17) to (20) with the three components of the equation of motion (1b) (with $i\sigma = s$ and in which we take the viscosity into account with the kinematic coefficient of viscosity η constant) constitute a system of 9 independent equations in 9 variables

$(\varrho', p', T', \bar{\mu}', X'_1, X'_3, \delta x, \delta y, \delta z).$

If the space dependence is of the form

$$f' = f e^{i(k_x x + k_y y + k_z z)}$$

introducing a new wave number k_z along the vertical axis z opposed to g, the compatibility condition, assuming all coefficients constant, yields the dispersion relation

$$s^4 + s^3(2\eta k^2 + \delta) + s^2 \left(\eta^2 k^4 - \frac{k_H^2}{k^2} g\mathscr{A} + 2\eta k^2 \delta \right) +$$

$$s \left[\frac{k_H^2}{k^2} g \delta_1 \left(\frac{\mu_1}{X_1} \frac{dX_1}{dz} + \frac{\mu_3}{X_3} \frac{dX_3}{dz} \right) + \frac{k_H^2}{k^2} g \frac{1}{\bar{\mu}} \frac{d\bar{\mu}}{dz} \delta + \right.$$

$$\left. + \eta^2 k^4 \delta - \eta k_H^2 g\mathscr{A} \right] +$$

$$+ \left[\eta k_H^2 g \delta_1 \left(\frac{\mu_1}{X_1} \frac{dX_1}{dz} + \frac{\mu_3}{X_3} \frac{dX_3}{dz} \right) + \eta k_H^2 g \frac{1}{\bar{\mu}} \frac{d\bar{\mu}}{dz} \delta_2 \right] = 0, \quad (21)$$

where

$$k^2 = k_x^2 + k_y^2 + k_z^2, \quad k_H^2 = k_x^2 + k_y^2$$

$$\delta = \frac{4-3\beta}{\beta} \left(\frac{\varepsilon}{C_p T} - m \right) - \left(\frac{v\varepsilon}{C_p T} - \frac{Kk^2}{C_p \varrho} + n \right) = \delta_1 - m \frac{4-3\beta}{\beta} - \delta_2$$

with

$$m = \frac{1}{C_p T} \left(\frac{K}{\varrho} - K_\varrho \right) \frac{d^2 T}{dz^2} + \frac{1}{C_p T} \frac{dT}{dz} \left(\frac{1}{\varrho} \frac{dK}{dz} - \frac{dK_\varrho}{dz} - ik_z K_\varrho \right)$$

$$n = \frac{1}{C_p T} K_T \frac{d^2 T}{dz^2} + \frac{1}{C_p \varrho} \frac{dT}{dz} \left[\frac{dK_T}{dz} + ik K_T \right] + \frac{ik_z}{C_p \varrho} \frac{dK}{dz}$$

$$K_\varrho = \left(\frac{\partial K}{\partial \varrho} \right)_T, \quad K_T = \left(\frac{\partial K}{\partial T} \right)_\varrho$$

If the chemical composition is constant and all non-conservative terms negligible, (21) reduces to

$$s^2 = \frac{k_H^2}{k^2} g\mathscr{A}$$

which shows that dynamical instability leading to proper convection appears only if $\mathscr{A} > 0$ and that, in this simple case, the latter is the stronger the higher the ratio of the vertical to the horizontal wavelengths.

If we neglect changes in chemical composition, we are left with

$$s^3 + s^2(2\eta k^2 + \delta) + s \left(\eta^2 k^4 - \frac{k_H^2}{k^2} g\mathscr{A} + 2\eta k^2 \delta \right) + (\eta^2 k^4 \delta - \eta k_H^2 g\mathscr{A}) = 0$$

which to the first order in the small quantities δ, ηk^2 reduces, apart from a trivial secular root, to the problem studied by Defouw (1970) with his $\mathscr{L}_T = -\nu\varepsilon/T$, $\mathscr{L}_\varrho = -\varepsilon/\varrho$ and $m = n = 0$, $\beta = 1$; of course, without rotation or magnetic field.

If, on the other hand, we neglect the energy generation ($\varepsilon = 0$) and viscosity ($\eta = 0$) and treat dT/dz and K/ϱ as constant in perturbing $(1/\varrho \text{ div} \mathbf{F})$ (i.e. $m = 0$, $n = 0$), we obtain

$$s^3 + s^2 \frac{Kk^2}{C_p\varrho} - s\frac{k_H^2}{k^2} g\mathscr{A} - k_H^2 g \frac{K}{C_p\varrho} \frac{1}{\bar{\mu}} \frac{d\bar{\mu}}{dz} = 0$$

which is Kato's equation (1966). Ignoring again the secular root and keeping only first order terms, this reduces to:

$$s^2 + \frac{Kk^2}{C_p\varrho} \frac{4-3\beta}{\beta} \frac{\mathscr{S}}{\mathscr{A}} s - \frac{k_H^2}{k^2} g\mathscr{A} = 0$$

with roots

$$s_{1,2} = -\frac{Kk^2}{2C_p\varrho} \frac{4-3\beta}{\beta} \frac{\mathscr{S}}{\mathscr{A}} \pm \sqrt{\frac{k_H^2}{k^2} g\mathscr{A}}.$$

It can be seen immediately that if $\mathscr{A} < 0$ but $\mathscr{S} > 0$, i.e. if condition (16) is satisfied, there are oscillating solutions with increasing amplitudes. It might be interesting to generalize the discussion to include the effects of viscosity and let K depend on ϱ and T but the corresponding dispersion equation which is of the 4th degree is harder to handle.

Of course, in this context, a local instability is only an indication and in any definite situation one should have to check whether it can actually give rise to a global instability for the star or whether stabilizing factors in other regions can overcome the local instability (cf. for instance Gabriel, 1969).

Finally, in the same conditions as above, but if we keep the terms in ε we get Dilke and Gough's (1972) equation

$$s^3 + s^2 \left(\frac{\varepsilon}{C_pT} - \frac{\nu\varepsilon}{C_pT} + \frac{Kk^2}{C_p\varrho}\right) - s\frac{k_H^2}{k^2} g\mathscr{A} +$$
$$+ \frac{k_H^2}{k^2} g \frac{\varepsilon}{C_pT} \left(\frac{\mu_1}{X_1} \frac{dX_1}{dz} + \frac{\mu_3}{X_3} \frac{dX_3}{dz}\right) + \frac{k_H^2}{k^2} g \frac{1}{\bar{\mu}} \frac{d\bar{\mu}}{dz} \left(\frac{\nu\varepsilon}{C_pT} - \frac{Kk^2}{C_p\varrho}\right) = 0$$

which, again discarding the secular root, has a solution which can be written in a first approximation

$$s_{1,2} = \pm \sqrt{\frac{k_H^2}{k^2} g\mathscr{A}} -$$
$$-\frac{1}{2}\left[\frac{\varepsilon}{C_pT}\left(1 + \frac{1}{\mathscr{A}}\left(\frac{\mu_1}{X_1}\frac{dX_1}{dz} + \frac{\mu_3}{X_3}\frac{dX_3}{dz}\right)\right) - \frac{\mathscr{S}}{\mathscr{A}}\left(\frac{\nu\varepsilon}{C_pT} - \frac{Kk^2}{C_p\varrho}\right)\right].$$

If $\mathscr{A} < 0$, we can indeed have amplified oscillating motions provided the square bracket is negative and abundances X_1 and X_3 increasing with z tend certainly to make this easier. If also $\mathscr{S} < 0$ which is probably the case in the Sun, the term in $\nu\varepsilon$ is also destabilizing while the effect of Kato's term is now in the opposite direction. Dilke and Gough estimated that, at least locally, modes could be unstable on this basis. Of course, this does not yet, of itself, prove that this would lead to an efficient mixing (cf. Ulrich and Rood, 1973).

On the other hand, as already pointed out above, only a global analysis can confirm whether or not the local tendency to instability is sufficiently strong to overcome the stabilizing influence of the surrounding layers which implies solving the problem of the vibrational stability of g^+ mode in a realistic solar model. Very recently Dziembowski and Sienkievicz (1973) have evaluated the total σ' as defined by (15) for models in the vicinity of the Sun. Neglecting the external convection zone, they find that all the g^+ modes should be stable. In particular, the g_1^+ mode which is the less stable with a period of the order of one hour has a damping time of the order of a few million years. A similar investigation has been made in Liège (Boury et al., 1974) and the results confirm those of Dziembowski and Sienkievicz yielding, if one stops the integral for σ' at the bottom of the convective envelope, an even somewhat shorter damping time. However if this integral, taking still into account only the radiative flux, is carried on close to the surface through the convection zone, the results is reversed and σ' becomes negative corresponding to an e-folding time of the order of 10^6 yr. Of course this is probably not significant because the convective flux and its variations including the horizontal components should be taken into account which is not too easy and a better approach should be used above the level where the quasi-adiabatic approximation breaks down. This work is now in progress in Liège. In any case, if this instability would persist it would be only slightly related to the Dilke and Gough mechanism and would be operative continuously and not only at definite intervals depending on the building up of gradients of X_1 and X_3.*

6. Variable Stars

As we have already recalled in Section (3.6), the β Canis Majoris stars still remain among the best candidates for an explanation in terms of non-radial pulsations and we have summarized there the new interesting suggestion of Osaki (1974) as to a possible origin of the latter. If, according to this idea, they result from a coupling through rotation with some overstable g^- modes it is likely that there would be appreciable energy from the buoyancy to drive them on and we would not have to depend on vibrational instability to excite them. Nevertheless, it would be very interesting also to compute the appropriate σ' to see how much of the original available energy would be dissipated and whether a balance at a reasonable amplitude can be reached. A somewhat similar idea (Ledoux, 1967) but applied this time to a narrow external convective zone could still be useful in the discussion of magnetic variables.

The characteristics of a recently discovered new type of variables, the so-called

---

* **Note added in proof.** Since then, early evolutionary solar models (2.4×10^8 to 3×10^9 yr) have been found unstable (Boury et al., 1974; Christensen-Dalsgaard and Gough, 1974), but the 'present' Sun, with a better treatment of convection, is stable.

white-dwarf variables, also strongly suggest non-radial pulsations. The first one DQ Her, was discovered by Walker (1956) and with a period of 71 seconds it remained the shortest period variable known for quite a long time. Its recent study by Warner et al. (1972) has brought direct support to the non-radial oscillation hypothesis by showing that the observed phase variations of the pulsation during eclipse are difficult to explain except in terms of a grazing eclipse of a white dwarf oscillating in a non-radial quadrupole ($l=2$) mode. Since then, Warner finds that agreement is even improved if the oscillation is not axisymetric but corresponds to the mode $l=2, m=2$ and another star UX Ma has been shown to have similar characteristics.

In the last few years, thanks mainly to the work of Warner, many stars have been added, some even with shorter periods (cf. Table V). For instance Warner and Robinson (1972) reported oscillations with periods near 17, 24 and 31 s respectively during outbursts of the dwarf novae Z Cam, CN Ori and AH Her and one with a period of 29 s in UX Ma. More recently Warner and Harwood (1973) added another of these fast pulsators to the list as they found brightness variations of considerable amplitude with a period of 28.15 s in another dwarf nova VW Hyi while it was declining from its recent outburst.

The list was also extended towards longer periods, first with T Cor Bor with a period of 98.2 s (Lawrence et al., 1967) and then a whole group cited here by order of increasing periods: G 61-29:105 s (Richer et al., 1973), HZ 29 (Am C Vn):115 s (Warner and Robinson, 1972), R 548:213, 273 s (Lasker and Hesser, 1971b), G 44-32: 600 s (Lasker and Hesser, 1971a), HL Tau: 746 s (Landolt, 1968; Warner and Robinson, 1972). But this last star was recognized as multi-periodic and Fitch (1973) has given a very nice analysis which isolates another period of 494 s coupled to the first and both modulated by a long period of 3.24 h. Still other candidates have been presented like BD 14341 with a period of 840 s (Williams, 1966) and SS Cyg, 960 s (Zuckerman, 1961) where the pulsation, however, has a definitely transient character although this may be true of a large fraction of these objects.

On the one hand, such short periods can only arise in small dense stars having either reached the white-dwarf stage or well on the way to it. But, on the other hand, the work on radial pulsations of such stars (cf. for instance Vila, 1970; Van Horn et al., 1972) shows that the periods in this case are much shorter than the observed ones except perhaps the very shortest ones. The same is true of the f and p modes of non-radial oscillations but the g modes offer interesting possibilities, as far as longer periods are concerned especially in white dwarfs. In fact the structure of a classical cold white dwarf completely degenerate is barotropic and $\mathscr{A} \to 0$ so that the periods of all g modes tend to infinity.

In realistic white dwarfs models $\mathscr{A}$ will no longer be zero. While still fairly small in the degenerate interior, it will become negative and appreciable in an external radiative envelope until in the very thin outer convective layers (Böhm, 1968; Van Horn, 1970), it tends towards zero again or to a very small positive value. The region with $\mathscr{A}$ negative and relatively large in absolute value will be the more extensive and the g periods the shorter the hotter the white dwarfs. Thus one may expect the periods of g modes to

TABLE V

Some short-period variables

Star	Nature	Period	Remarks
Z Cam	dw. nova	17 s	
CN Ori	dw. nova	24 s	
VW Hyi		28.1 s	
UX U Ma	DA	29 s	Eclipsing binary 4^{h}43^m
AH Her	dw. nova	31 s	
DQ Her	nova	71 s	Eclipsing binary 4^{h}39^m
T Cor Bor		98.2 s	
G 61–29	DB	105 s	Eclipsing binary 6^{h}16^m
HZ-29 (AM CVn)	DB$_p$	115 s 1015 s (orbital?)	
R 548	DA	213 s 273 s	
G 44–32	DC	600 s 822 s 1638 s (orbital?)	
HL Taur	DA	746 s 494 s Long mod. (3.24 h)	

increase as the star cools. All this was already apparent from the discussion by Baglin and Schatzman (1969) and from the computations of Harper and Rose (1970) or from general approximate expressions of the σ_{g+}^2 as given in Ledoux and Walraven (1958, Section 79). This was essentially the basis for the proposed interpretation by Chanmugam (1972) and by Warner and Robinson (1972) in terms of g^+ modes. Multiple periodicities and the direct evidence of DH Her and UX Ma during eclipses added new arguments.

The theoretical expectations have now been confirmed in details by Osaki and Hansen (1973) who have studied the non-radial oscillations of various cooling white dwarfs models for two masses: 0.398 and 1 $M_\odot$. They find, for instance, that the period $P_{g_1^+}$ of the g_1^+ mode varies, in the first case, from 49.02 s to 209.8 s as R varies from 4.14×10^{-2} to 1.39×10^{-2} $R_\odot$ while, for the higher mass, $P_{g_1^+}$ varies from 8.15 to 111.9 s as R varies from 1.12×10^{-2} to 6×10^{-3} $R_\odot$. The lower range of observed periods is thus reasonably covered. However for the longer periods observed, one would either have to go to rather unrealistic high g^+ modes or find some factor in the structure of the star which could reduce $\mathscr{A}$ again. Osaki and Hansen suggest that the external convection zone ($\mathscr{A} \simeq 0$) which they have neglected could play this rôle but it would have to extend fairly deep in order to bring agreement with the longest observed periods. In fact, in collaboration with Böhm et al. (1973) we have also computed the periods of various non-radial modes for a $0.6\,M_\odot$ white dwarf model with a fitted outer convection zone and $T_{\text{eff}} = 12\,000$ K. The periods found are reproduced in Table VI as well as the Q values. The jump in period and Q value in going from the p and f modes to the g modes is particularly well marked in this model which is somewhat similar in this respect to the coolest models of Osaki and Hansen. However here the external convection zone is so thin that, although it reduces the increase of the amplitude of the g_1 mode somewhat towards the surface, it has a relatively small influence on its period.

TABLE VI

Periods (in sec) of various modes of non-radial oscillations for a white dwarf model of $0.6\,M_\odot$, $T_{\text{eff}} = 12\,000$ K and a narrow outer convection zone

Modes	p_4	p_3	p_2	p_1	f	g_1	g_2	g_3	g_4	g_5	g_6
Period	2.38	2.93	3.85	5.73	11.9	200	250	347	423	528	646
Q	0.0483	0.0594	0.0761	0.1162	0.2414	4.057	5.071	7.039	8.581	10.711	13.105

As Osaki and Hansen point out, since most of these variables are novae or dwarf novae or members of close binaries, the excitation of the non-radial modes need not necessarily correspond to an intrinsic vibrational instability since they could be excited by their outbursts or by tidal action. Nevertheless, they computed the damping coefficient σ' and since they assumed no nuclear energy generation they found only positive σ' corresponding to damping times due to radiative transfer and neutrino emission of the order of 10^{11} yr for the f mode of the smaller mass ($0.4\,M_\odot$) to values as small as 10^5 and 3000 yr for the g_1 mode of respectively the $0.4\,M_\odot$ and $1.0\,M_\odot$ star. While the longest damping time is of the same order as for radial modes (Van Horn et al., 1972), the damping times of the g modes are much shorter. This can be qualitatively understood since these g modes have appreciable amplitudes only in the external part of the white dwarf (cf. Figure 11) which contains relatively little mass, so that the total energy of the pulsations K is relatively small. On the other hand, this is the main region of conductive dissipation so that $(\Delta K)_P$ per period is relatively large and $\sigma' = -(1/2)(\Delta K)_P/K$ is large.

Another factor which is not a priori negligible for these dense stars is the emission of gravitational waves studied first by Thorne and Campolattaro (1967), Thorne (1969),

Ipser and Thorne (1973). Osaki and Hansen (1973) show that indeed this can reduce drastically the damping time of the p modes which, for the 1 $M_\odot$ star, can then become as small as 10 to 100 yr but hardly affects the g modes.

On the other hand, as stressed by Warner (1972), it is likely that outbursts in all the cataclysmic variables originate in the white dwarf components in the form of non-radial modes corresponding to spherical harmonics of various degrees. Is this connected as suggested by Rose (1968), Rose and Smith (1972) to vibrational instability following a thermal runaway at the surface of the degenerate core of an accreting

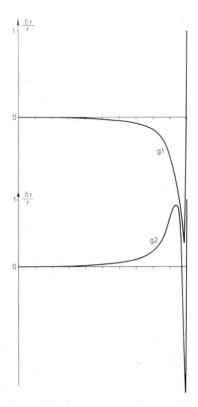

Fig. 11. Run of $\delta r/r$ for g_1 and g_2 modes ($l=2$) in a white dwarf of $0.6M_\odot$ with a narrow outer convection zone.

white dwarf? In fact Rose found vibrational instability for radial modes and his paper with Harper (1970) was in part aimed at discussing the secondary excitation of g modes by coupling. But with the distribution of the amplitudes of g modes in white dwarfs as described above, it is very likely that these modes could themselves become easily and strongly vibrationally unstable in the case considered by Rose. In our work with Böhm and Robe, we want to check this and preliminary calculations show that indeed very little hydrogen in the external layers generating only a small fraction of the energy radiated would be sufficient to lead to vibrational instability for the g modes.

It is clear however that these rapid pulsators with often many close periodicities and their rapidly changing periods offer many other interesting and puzzling problems.

We cannot here, even summarly, go into the problem of pulsars and variable X-ray sources but non-radial pulsations may still be relevant and, in any case, they may play an important rôle in the final dense stages of evolution by the emission of gravitational radiation, especially through p modes (cf. Hartle *et al.*, 1972).

In ending, let us draw the attention again to the general possibility of variability in binaries. In a recent paper, Herbst (1973) has reviewed this problem in binaries with strong Ca II emission and has added at least two new cases. The light period does not always agree exactly with the orbital period but is often of the same order and they certainly set a nice new problem to the theorists.

References

Baglin, A. and Schatzman, E.: 1969, in S. S. Kumar (ed.), *Low Luminosity Stars*, Gordon and Breach, New York, p. 385.
Böhm, K.-H.: 1968, *Astrophys. Space Sci.* **2**, 375; also 1969, in S. S. Kumar (ed.), *Low Luminosity Stars*, Gordon and Breach, New York, p. 393.
Böhm, K.-H., Ledoux, P., and Robe, H.: 1973 (in preparation).
Bolt, B. A. and Derr, J. S.: 1969, *Vistas in Astronomy* **11**, 69.
Boury, A., Gabriel, M., Ledoux, P., Noels, A., and Scuflaire, R.: 1974, XIXe Coll. Astrophys. Liège, *Mem. Soc. Roy. Sci. Liège* (in press).
Camps, J.: 1973, Mémoire de Licence, Sci. Math. Université de Liège.
Chandrasekhar, S.: 1961, *Hydrodynamic and Hydromagnetic Stability*, Clarendon Press, Oxford.
Chandrasekhar, S.: 1964, *Astrophys. J.* **139**, 664.
Chandrasekhar, S.: 1969, *Ellipsoidal Figures of Equilibrium*, Yale University Press, New Haven and London.
Chandrasekhar, S. and Lebovitz, N.: 1964, *Astrophys. J.* **140**, 1517.
Chanmugam, G.: 1972, *Nature Phys. Sci.* **236**, 83.
Christensen-Dalsgaard, J. and Gough, D. O.: 1974, XIXe Coll. Astrophys. Liège, *Mem. Soc. Roy. Sci. Liège* (in press).
Counson, J., Ledoux, P., and Simon, R.: 1956, *Bull. Soc. Roy. Sci. Liège* **36**, 144.
Cowling, T. G.: 1941, *Monthly Notices Roy. Astron. Soc.* **101**, 367.
Cowling, T. G. and Newing, R. A.: 1949, *Astrophys. J.* **109**, 149.
Defouw, R. J.: 1970, *Astrophys. J.* **160**, 659.
Dilke, F. W. W. and Gough, D. O.: 1972, *Nature* **240**, 262.
Dziembowski, W.: 1971, *Acta Astron.* **21**, 289.
Dziembowski, W. and Sienkievicz, R.: 1973, Repr. No. 30, Inst. Astron. Warszawa.
Eckart, C.: 1960, *Hydrodynamics of Oceans and Atmospheres*, Pergamon Press, London, New York.
Einsenfeld, J.: 1968a, *J. Math. Anal. Appl.* **23**, 58.
Einsenfeld, J.: 1968b, *J. Math. Anal. Appl.* **26**, 357.
Ezer, D. and Cameron, A. G. W.: 1972, *Nature Phys. Sci.* **240**, 180.
Fitch, W. S.: 1973, *Astrophys. J. Letters* **181**, L95.
Fowler, W. A.: 1972, *Nature* **238**, 24.
Fowley, W. M.: 1972, *Astrophys. J.* **180**, 483.
Gabriel, M.: 1969, *Astron. Astrophys.* **1**, 321.
Goosens, M. and Smeyers, P.: 1974, *Astrophys. Space Sci.* **26**, 137.
Harper, R. V. R. and Rose, W. K.: 1970, *Astrophys. J.* **162**, 963.
Hartle, J. B., Thorne, K. S., and Chitre, S. M.: 1972, *Astrophys. J.* **176**, 177.
Herbst, W.: 1973, *Astron. Astrophys.* **26**, 137.

Hitotuyanagi, Z. and Takeuti, M.: 1964, Sandai Astron. Rept. No. 88.
Hurley, M., Roberts, P. H., and Wright, K.: 1966, *Astrophys. J.* **143**, 535.
Ince, E. L.: 1964, *Ordinary Differential Equations*, Dover Publ., New York.
Ipser, J. R. and Thorne, K. S.: 1973, *Astrophys. J.* **181**, 181.
Iweins, P. and Ledoux, P.: 1971 (unpublished).
Kato, S.: 1966, *Publ. Astron. Soc. Japan* **18**, 374.
Kopal, Z.: 1949, *Astrophys. J.* **109**, 509.
Landolt, A. U.: 1968, *Astrophys. J.* **153**, 151.
Langer, R. E.: 1959, *Trans. Am. Math. Soc.* **90**, 113.
Lasker, B. M. and Hesser, J. E.: 1971a, *Astrophys. J. Letters* **163**, L89.
Lasker, B. M. and Hesser, J. E.: 1971b, *Astrophys. J. Letters* **158**, L171.
Laurence, G. M., Ostriker, J. P., and Hesser, J. E.: 1967, *Astrophys. J. Letters* **148**, L161.
Lebovitz, N. R.: 1965, *Astrophys. J.* **142**, 229.
Lebovitz, N. R.: 1966, *Astrophys. J.* **146**, 946.
Ledoux, P.: 1949, *Mém. Soc. Roy. Sci. Liège* **9**, Contribution à l'étude de la structure interne des étoiles et de leur stabilité, Chapters IV and V.
Ledoux, P.: 1951, *Astrophys. J.* **114**, 373.
Ledoux, P.: 1967, in R. C. Cameron (ed.), *The Magnetic and Related Stars*, Mono Book, Baltimore, p. 65.
Ledoux, P.: 1969, 'Oscillations et Stabilité Stellaire' in *La Structure interne des étoiles*, XIe Cours de Perfectionnement de l'Association Vaudoise des Chercheurs en Physique, Saas-Fee, 24–29 mars 1969.
Ledoux, P. and Walraven, Th.: 1958, *Handbuch der Physik* **51**, Chapter IV.
Ledoux, P. and Smeyers, P.: 1966, *Compt. Rend. Acad. Sci. Paris*, Sér. B, **262**, 841.
Leighton, R. B.: 1961, 4th IAU-IUTAM Symposium on Cosmical Dynamics: *Aerodynamical Phenomena in Stellar Atmospheres, IAU Symp.* **12**, Suppl. *Nuovo Cimento* **22**, Sér. X, p. 321.
McKenzie, J. F.: 1971, *Astron. Astrophys.* **15**, 450.
Moore, D. W. and Spiegel, E. A.: 1966, *Astrophys. J.* **143**, 871.
Noels, A., Boury, A., Scuflaire, R., and Gabriel, M.: 1974, submitted to *Astron. Astrophys.*
Osaki, Y.: 1971, *Publ. Astron. Soc. Japan* **23**, 485.
Osaki, Y.: 1974, *Astrophys. J.* **189**, 469.
Osaki, Y. and Hansen, C. J.: 1973, *Astrophys. J.* **185**, 277.
Owen, J. W.: 1957, *Monthly Notices Roy. Astron. Soc.* **117**, 384.
Pekeris, C. L.: 1938, *Astrophys. J.* **88**, 189.
Perdang, J.: 1968, *Astrophys. Space Sci.* **1**, 355.
Richer, H. B., Auman, J. R., Isherwood, B. C., Steele, J. P., and Ulrych, T. J.: 1973, *Astrophys. J.* **180**, 107.
Robe, H.: 1965, *Bull. Classe Sci. Acad. Roy. Belg.*, 5e Sér., **51**, 598.
Robe, H.: 1968, *Ann. Astrophys.* **31**, 475.
Robe, H.: 1973 (private communication).
Robe, H.: 1974, *Bull. Soc. Roy. Sci. Liège* **18**, 240.
Robe, H. and Brandt, L.: 1966, *Ann. Astrophys.* **29**, 571.
Robe, H., Ledoux, P., and Noels, A.: 1972, *Astron. Astrophys.* **18**, 424.
Rood, R. T.: 1973, *Nature Phys. Sci.* **240**, 178.
Rose, W. K.: 1968, *Astrophys. J.* **152**, 245.
Rose, W. K. and Smith, R. L.: 1972, *Astrophys. J.* **172**, 699.
Saslaw, W. S. and Schwarzschild, M.: 1965, *Astrophys. J.* **142**, 1468.
Schatzman, E. and Soufrin, P.: 1967, *Ann. Rev. Astron. Astrophys.* **5**, 67.
Scuflaire, R.: 1973 (private communication).
Simon, R.: 1957, *Bull. Classe Sci. Acad. Roy. Belg.*, 5e Sér., **43**, 610.
Smeyers, P.: 1966, *Ann. Astrophys.* **29**, 539.
Smeyers, P.: 1967, *Bull. Soc. Roy. Sci. Liège* **36**, 357.
Smeyers, P.: 1968, *Ann. Astrophys.* **31**, 159.
Soufrin, P. and Spiegel, E. A.: 1967, *Ann. Astrophys.* **30**, 985.
Spiegel, E.: 1964, *Astrophys. J.* **139**, 959.
Tassoul, J.-L.: 1967, *Ann. Astrophys.* **30**, 363.
Tassoul, M. and Tassoul, J.-L.: 1968, *Ann. Astrophys.* **31**, 251.
Thomas, J. H., Clark, P. A., and Clark, A., Jr.: 1971, *Solar Phys.* **16**, 51.
Thomson, W.: 1863, *Phil. Trans. Roy. Soc. London* **153**, 612.
Thorne, K. S.: 1969, *Astrophys. J.* **158**, 1.

Thorne, K. S. and Campolattaro, A.: 1967, *Astrophys. J.* **149**, 591.
Tolstoy, I.: 1963, *Rev. Mod. Phys.* **35**, 207.
Ulrich, R. K.: 1970, *Astrophys. J.* **162**, 993.
Ulrich, R. K. and Rood, R. T.: 1973, *Nature Phys. Sci.* **241**, 111.
Vandakurow, Y. V.: 1967, *Astron. Zh.* **44**, 786.
Van der Borght, R. and Wan Fook Sun, 1965, *Bull. Classe Sci. Acad. Roy. Belg.* 5e Sér., **51**, 978.
Van Horn, H. M.: 1970, *Astrophys. J. Letters* **160**, L53.
Van Horn, H. M., Richardson, M. B., and Hansen, C. J.: 1972, *Astrophys. J.* **172**, 181.
Vila, S. C.: 1970, *Astrophys. J.* **162**, 971.
Walker, M. F.: 1956, *Astrophys. J.* **123**, 68.
Wan, F. S.: 1966, Ph. D. Thesis, Australian National University.
Warner, B.: 1972, *Monthly Notices Roy. Astron. Soc.* **160** (Short comm.), 35 p.
Warner, B. and Harwood, J. M.: 1973, Commission 27, *IAU Information Bulletin* No. 756.
Warner, B., Peters, W. L. Hubbard, W. B., and Nather, R. E.: 1972, *Monthly Notices Roy. Astron. Soc.* **159**, 321.
Weinberger, H. F.: 1968, *J. Math. Anal. Appl.* **21**, 506.
Williams, J. O.: 1966, *Publ. Astron. Soc. Pacific* **78**, 279.
Wolff, C. L.: 1972a, *Astrophys. J.* **176**, 833.
Wolff, C. L.: 1972b, *Astrophys. J. Letters* **177**, L87.
Zahn, J. P.: 1968, *Astrophys. Letters* **1**, 209.
Zuckerman, M.-C.: 1961, *Ann. Astrophys.* **24**, 431.

DISCUSSION

Warner: Have you computed a C_L for a white dwarf yet?

Ledoux: I have not done it myself, no.

Warner: Do you have any feeling for what it could be? It is 0.15 for your calculation for a main sequence star.

Ledoux: I have no idea. I wouldn't like to give any number.

Tayler: It is a different problem asking whether instability will occur and discussing what happens once it is developed. As to the penetration question, do you think that with a linear treatment you get really useful information on what will happen once the convection is developed from this or just the very crude first idea?

Ledoux: I think you get only a crude first idea. Of course, what you do is to evaluate how much the gradient has to be superadiabatic in the turbulent picture, to transfer the heat. Then on that basis you compute g modes for that case and you find out how far they can penetrate the stable region. It certainly is not going to give you very precise results but at least some indications, probably as good as by any other approach.

Maeder: What typical characteristic of the light curve might result from the presence of some non-radial pulsation in the star?

Ledoux: In general non-radial pulsation would contribute very little to the light curves because you have parts of the star which are brighter, parts which are darker, and so on. So it is not a very efficient way of producing light variation, but there will be some. It will depend on the detail, which mode is excited and so on. If you go to high modes, of course, you would not see anything because you would have spots on the star, light and dark spots. But for $l=2$ for example, you would have a region which is bright and a region which is dark. This will combine with rotation too and it may be quite complicated to compute the light curve.

Buscombe: I think we do not know nearly enough about the rotational speed of stars in which there are other instabilities. I would make a plea to observers to work on this a bit because I feel that this is a parameter which is turning out to be very significant about stars. For example, in the MK classification, without detracting the great success that it has had, I think that this is completely suppressed and that we need this additional information. Part of this is confused with the macro turbulence. I think with such devices as the Fabry-Perot interferometer, there are possibilities.

THE SUN AS A PULSATING ROTATING STAR*

KENNETH H. SCHATTEN

Victoria University of Wellington, Wellington, New Zealand

Abstract. Physical arguments are provided which suggest the following:

(a) The Sun rotates rapidly internally, with a period near one day. The arguments are based upon a low 'effective' plasma *dynamic* viscosity associated with a negative 'effective' magnetic density, $\varrho_M = -B^2/4\pi v^2$. This low (near zero) viscosity allows several calculations of the angular velocity of the solar core to be made. A fluid dynamical argument based upon the inviscous Navier-Stokes relation shows that for objects seated in a *non-expanding* magnetohydrodynamic fluid, the usual Kepler law should be replaced by $T^2 \propto r^4$.

(b) The rapid rotation suggests that the solar magnetic field is deeply buried and, furthermore, that the Sun violates the Ferraro theorem. This violation results from a radial electric field necessary to support the solar plasma. This radial field requires the Sun to be charged with a charge $q = +2 \times 10^{11}$ esu. Thus the Sun's differential rotation may be viewed as arising from $B']_\theta = 0$ in the rest frame of the fluid where $\mathbf{E}_r \neq 0$ or alternatively as a polar spin down by the solar wind flow. Thus in the fluid frame, the solar activity cycle may be viewed as an Alfvén wave, with $T = \lambda/v_A = 20$ yr. The velocity of material along the field is such that $\varrho v/B = \text{const} \cong 4 \times 10^{-6}$ gm (G^{-1} cm^{-2} s^{-1}), governed by the collapse of mass in the solar core; $(4H + 4e \rightarrow 1He + + 2e)$ results in a reduction of gas pressure unless $\sim 7 \times 10^{14}$ gm s^{-1} of material continually collapse so as to conserve the particle number in the solar interior.

(c) This requires the Sun to emit gravity waves with an energy comparable to its luminosity. The *virial* of the nuclear reactions in the core governs whether the energy goes into *gravity waves* or *heat*. A variable positron-electron core allows this.

---

* Due to lack of time, only the first part of this paper was presented.

INSTABILITIES OF MAGNETIC FIELDS IN STARS

R. J. TAYLER

Astronomy Centre, University of Sussex, England

Abstract. It has been shown (Markey and Tayler, 1973; Tayler, 1973; Wright, 1973) that a wide range of simple magnetic field configurations in stars are unstable. Although the ultimate effect of the instabilities is unclear, it seems likely that they would lead to enhanced destruction of magnetic flux, so that magnetic field decay would be much more rapid than previously supposed. Instability is almost certain in a non-rotating star containing either a purely toroidal field or a purely poloidal field, which has closed field lines inside the star. In both cases the instability resembles the well known instabilities of cylindrical and toroidal current channels, modified by the constraint that motion must be almost entirely along surfaces of constant gravitational potential.

If both toroidal and poloidal fields are present, the problem is more complicated. In a toroidal plasma with a helical field, the worst instabilities are also helical but it is impossible for a helical disturbance to be parallel to a surface of constant gravitational potential everywhere. As a result, the admixture of toroidal and poloidal fields has a stabilizing influence, but it is not at present clear whether the majority of such configurations are completely stable.

The effect of rotation has not yet been studied but it will certainly be important if the rotation period is less than the time taken for an Alfvén wave to cross the region of interest. This is true in most stars unless the internal magnetic field is very much stronger than any observed field.

References

Markey, P. and Tayler, R. J.: 1973, *Monthly Notices Roy. Astron. Soc.* **163**, 77.
Tayler, R. J.: 1973, *Monthly Notices Roy. Astron. Soc.* **161**, 365.
Wright, G. A. E.: 1973, *Monthly Notices Roy. Astron. Soc.* **162**, 339.

DISCUSSION

Schatten: I would like to ask either of the last two speakers whether or not they could come up with a complete computer model for their proposed mechanisms for creating fields in the solar core, because one can either neglect magnetic fields and do a complete computer analysis which is incorrect or one can include magnetic fields and must resort to some sort of physical argument.

Tayler: Sorry, I haven't quite understood whether that was a question or a statement.

Schatten: A question. Whether or not you can actually provide calculations which support your views.

Tayler: Oh. The calculations of the existence of the instability are complete. The calculations of the development of the instability are non-existent, but linear calculations of the existence of the instability are complete.

MAGNETIC FIELDS IN X-RAY BINARY SYSTEMS

JAMES C. KEMP*

Institute for Astronomy, University of Hawaii, Honolulu, H.I. 96822, U.S.A.

Abstract. The case for variable magnetic fields, in the range 1–15 kG, in three of the X-ray binary candidate stars, is discussed. We refer to electronic Zeeman polarimetry by Kemp and Wolstencroft (1973a and b) and by Angel *et al.* (1973). Our detections in Hβ showed: (1) definite magnetic fields, roughly constant over periods of 2 h, in θ^2 Orionis and HD 77581; and (2) complex time behavior in HD 153919, with field reversals over periods of $\sim$ 10 min, but with usually small values ($\lesssim$ 2 kG) as averaged over times of 1–2 h. (Photographic Zeeman spectroscopy is therefore unlikely to show detectable fields, i.e. $>$ 2 kG, in the latter star.) The essentially null results of Angel *et al.* (1973) in Hα, in HD 77581 and HD 153919, indicate that the emission is non-magnetic; only the underlying absorption (which is far weaker than the emission in Hα) shows Zeeman effects. A plot of the Zeeman measure for HD 77581, against the 8.95-day period, is shown: our points are suggestive of periodicity; and the points of Angel *et al.* (1973) which unfortunately have larger relative error bars, are consistent with the indicated variation and with the general scale of our results.

References

Angel, J. R. P., McGraw, J. T., and Stockman, H. S. Jr.: 1973, *Astrophys. J. Letters* **184**, L79.
Kemp, J. C. and Wolstencroft, R. D.: 1973a, *Astrophys. J. Letters* **182**, L43.
Kemp, J. C. and Wolstencroft, R. D.: 1973b, *Astrophys. J. Letters* **185**, L21.

DISCUSSION

Buscombe: I would appeal to you to tackle some of supergiant A stars I was talking about yesterday, because I think you would see some activity of this kind in the Hα.
Kemp: We have.
Buscombe: What do you get?
Kemp: 25 Orionis. We see enormous effects, we haven't published them yet. That is a Be emission star; we have several more, δ Cas is another one. It's a rapid rotator. They're non-periodic, that's one reason we haven't said anything about them.
Buscombe: The lines are so broadened by turbulence that you'd have no hope with the conventional magnetic analyser.
Kemp: What we see is all kinds of fantastic time variations, linear polarization effects particularly in the Balmer lines.
Buscombe: You have problems with the depolarization in the interstellar medium because these are distant supergiants.
Kemp: Yes, in all these cases there is a large interstellar polarization which one has to deal with.

* Also Department of Physics, University of Oregon, Eugene, Oregon 97403.

Tayler: Do you have any estimate at all of the strength of the field you're looking at?

Kemp: Yes, I didn't say anything about this, and we were overly enthusiastic in assigning field strengths to these results in the first paper. It's very complicated because of the spectrum. However, these fields roughly correspond to 1–15 kG. It's very sensitive to this whole question of the emission and absorption components. You can either have a region where you see where the magnetic field is falling off in some sense and there is emission and you get a negative contribution. If you assume an equivalent width of about 1 Å in Hβ in emission, these correspond to fields around a couple of kG.

THE OSCILLATIONS AND THE STABILITY OF ROTATING MASSES WITH MAGNETIC FIELDS

R. K. KOCHHAR and S. K. TREHAN

Dept. of Mathematics, Panjab Univerlsty, Chandigarh, India

Abstract. It is shown that when a magnetic field is present along the axis of rotation, the point of bifurcation, where the Jacobi ellipsoids branch off from the Maclaurin spheroids, occurs at a value of eccentricity higher than the value $e = 0.81267$ that obtains in the absence of a magnetic field. This is in contrast with the effect of a toroidal magnetic field which, as has been shown earlier, leaves the point of bifurcation unaffected.

B. ROTATION – CLOSE BINARIES

ROTATIONAL INSTABILITIES AND STELLAR EVOLUTION

JEAN-PAUL ZAHN

Observatoire de Nice, Le Mont-Gros, 06300 – Nice, France

Abstract. This review deals with the local instabilities arising when the effects of rotation are taken into account in the evolution of a non-magnetic star.
The Rayleigh and shear instabilities will be examined under the conditions prevailing in radiative zones where the effect of density stratification, thermal diffusion, viscosity and varying chemical composition must be taken into account. The possible consequences on the evolution of a star are finally outlined.

In the past decade, many authors have described in detail the evolution of stars of various masses and chemical composition, but the vast majority of the calculated models apply to non-rotating, non-magnetic, and hence spherical stars. To introduce rotation or a magnetic field into the stellar models is not an easy task: the one-dimensional problem becomes at least two-dimensional, unless one settles for very drastic simplifications. More seriously, new dependent variables, namely the angular velocity and the components of the magnetic field, enter and one needs additional information, expressed by additional differential equations, about their behaviour in order to determine their distribution inside the star and their evolution with time.

For the sake of simplicity, the effects of rotation and magnetic field are most often studied separately, although one should bear in mind that they may be coupled in many instances. As suggested in the title above, we will deal here only with the instabilities which are triggered by the rotation of a non-magnetic star and which may influence its evolution. We shall narrow our field even further, and focus our attention on the *local* instabilities, as opposed to the large-scale Eddington-Sweet circulation or to the global instabilities arising when the rotational energy is an appreciable fraction of the total energy of the star. For these latter types of instabilities, we refer the reader to the reviews written by Mestel (1965, 1970) and by Ostriker (1970).

As for the local instabilities, the subject has been reviewed in the recent past by several authors, who also discussed their relevance to stellar structure and evolution (Lebowitz, 1967; Strittmatter, 1969; Spiegel and Zahn, 1970; Fricke and Kippenhahn, 1972; see also Ledoux, 1958). For the layman, such reviews may well appear as tiresome enumerations of all presently known instabilities, yet there could hardly be another approach to the problem since no criterion for *stability* has been found so far. In fact, a closer look at those seemingly impressive lists of local instabilities reveals that all derive from the well-known Rayleigh and shear instabilities, which are present even in an homogeneous incompressible fluid. Our purpose here will be to follow these classical instabilities under the conditions prevailing in the radiative zones of a star, taking into account such effects as density stratification, thermal diffusion, viscosity and varying chemical composition. Having done this, we will conclude with a sketch of the possible consequences of the described rotational instabilities on the evolution of a star.

1. Dynamical Instabilities

Dynamical instabilities occur at some place in the fluid if the equilibrium state is unstable towards an adiabatic inviscid perturbation. Such a dynamical instability has a local character and will settle the fluid into a new state of equilibrium. This new equilibrium state need not be a static one as it may display some motions, turbulent or not. If the equilibrium is dynamical, however, its description cannot be achieved in the adiabatic inviscid frame and one has to take into account both thermal diffusion and viscous transport of momentum.

What all dynamical instabilities have in common is their growth rate: they proceed on a dynamical time-scale which, in the case considered here, will be of the order of a rotation period.

1.1. Axisymmetric instability

The criterion for axisymmetric instability can be derived in many ways (see for instance Goldreich and Schubert, 1967; Fricke, 1968; James and Kahn, 1970), but in most cases some assumption is made about the spatial behaviour of the perturbation and the criterion obtained in such a way is then no longer rigourously a local one. Here, we prefer to follow the method introduced by Fjörtoft (Eliassen and Kleinschmidt, 1957).

Consider a non-magnetic inviscid star in static equilibrium, whose rotation law $\Omega(\varpi, z)$ depends both on the distance from the rotation axis, ϖ, and on the distance from the equatorial plane, z. Let us perturb this equilibrium state by a small axisymmetric and purely meridional displacement field $\delta \mathbf{r}(\varpi, z)$ and let us assume that the perturbation proceeds on a time scale short enough to keep it adiabatic. Since the fluid is in an equilibrium state, the first variation $\delta \varepsilon$ of the total (gravitational + internal + kinetic) energy ε is zero. Its second variation is given by the volume integral

$$\delta^2 \varepsilon = \int \left[\delta \mathbf{r} \cdot \mathfrak{M} \cdot \delta \mathbf{r} + \frac{1}{\gamma} \frac{P}{\varrho} \left(\frac{\delta P}{P} \right)^2 \right] \varrho \, d\tau, \tag{1}$$

where δP is the Eulerian pressure perturbation associated with the displacement $\delta \mathbf{r}$; P, ϱ, γ are the usual notations for pressure, density and the adiabatic exponent.

If the characteristic time of the perturbation exceeds the travel time of a sound wave across the test domain, the second term of this integrand will be negligible compared with the first one, whose sign will determine that of $\delta^2 \varepsilon$. If the quadratic form $\delta \mathbf{r} \cdot \mathfrak{M} \cdot \delta \mathbf{r}$ is positive definite, the stationary value of ε is a minimum and the tested equilibrium state is a stable one. On the other hand, if this form is not definite positive, it is possible to choose a virtual displacement $\delta \mathbf{r}$ which will decrease the total energy, thus proving that the equilibrium is then unstable.

This provides a genuine local instability criterion since the virtual displacement $\delta \mathbf{r}$ is entirely arbitrary – apart from the condition that it must vanish on the boundary of the integration domain and that it varies in a time scale which meets the requirements for both adiabaticy and pressure equilibrium to be achieved.

Now the tensor $\mathfrak{M}$ can be split into two parts, one representing the effect of the density stratification, the other that of the stratification of angular momentum:

$$\mathfrak{M} = \mathfrak{M}_1 + \mathfrak{M}_2 = \frac{1}{C_p}\operatorname{grad} S(-\mathbf{g}) + \frac{1}{\varpi^3}\operatorname{grad}(\varpi^2 \Omega)^2 \operatorname{grad} \varpi \qquad (2)$$

Here $\mathbf{g}$ is the local gravity (including the centrifugal acceleration), and S is the specific entropy:

$$\frac{1}{C_p}\operatorname{grad} S = \frac{1}{\gamma P}\operatorname{grad} P - \frac{1}{\varrho}\operatorname{grad}\varrho = \frac{1}{\varrho}\operatorname{grad}_{AD}\varrho - \frac{1}{\varrho}\operatorname{grad}\varrho.$$

(Note also that, when $\mathbf{g}$ and $\operatorname{grad} S$ are colinear or nearly so,

$$-\mathbf{g}\cdot\frac{1}{C_p}\operatorname{grad} S = \frac{|g|}{H_P}(\nabla_{AD} - \nabla_{RAD})$$

with the usual notations, H_P being the pressure scale-height). The discussion of the sign of the two quadratic terms $Q_1 = \boldsymbol{\delta r}\cdot\mathfrak{M}_1\cdot\boldsymbol{\delta r}$ and $Q_2 = \boldsymbol{\delta r}\cdot\mathfrak{M}_2\cdot\boldsymbol{\delta r}$ leads to a few well known instability criteria which we will review briefly.

If the fluid has a neutral density stratification ($\operatorname{grad} S = 0$), the first quadratic term is identically zero; instability occurs whenever Q_2 is not positive definite, i.e. when either of the following conditions is met

$$\frac{\partial}{\partial \varpi}(\varpi^2\Omega)^2 < 0; \quad \frac{\partial}{\partial z}(\varpi^2\Omega)^2 \neq 0. \qquad (3a, b)$$

The first of those conditions is the Rayleigh criterion for instability in a homogeneous inviscid and non gravitating fluid; the second need not be considered for such a fluid since the rotation law would then violate the Taylor-Proudman theorem and no equilibrium state would be possible (see Greenspan, 1968).

Let us now consider a star with a cylindrical rotation law $\Omega(\varpi)$ and ask if a stable density stratification may prevent the instability when the Rayleigh criterion predicts that it should occur. In such a star, the total body force (gravitational + centrifugal) derives from a potential and the surfaces of constant potential, pressure and density (and hence entropy in a homogeneous star) all coincide; the fluid is said to be barotropic. The quadratic form Q_1 is positive definite and, except at the exact equator of the star, Q_2 must also be positive definitive for the total quadratic form $Q_1 + Q_2$ to be of the same character. In other words, if the Rayleigh discriminant is negative, $(d/d\varpi)(\varpi^2\Omega)^2 < 0$, some displacements $\boldsymbol{\delta r}$ will decrease the total energy so that the equilibrium state is proved to be unstable. The unstable displacements are those for which $Q_1 = 0$ or nearly vanishes i.e. those which do not feel the density stratification because they are parallel to the equipotentials.

This result can be extended to more general rotation laws, for which Ω is a function of both ϖ and z. $\operatorname{grad} S$ and $\mathbf{g}$ are then no longer colinear and the angle between them is determined by the baroclinic condition

$$\frac{1}{C_P}\operatorname{grad} S \times (-\mathbf{g}) + \frac{1}{\varpi^3}\operatorname{grad}(\varpi^2\Omega)^2 \times \operatorname{grad}\varpi = 0 \qquad (4)$$

which is obtained by taking the curl of the equation of motion. The necessary and sufficient condition for the total quadratic form $Q_1 + Q_2$ to be positive definite is

$$\left[\frac{1}{C_P} \operatorname{grad} S \times \frac{1}{\varpi^3} \operatorname{grad}(\varpi^2 \Omega)^2\right] \cdot [-\mathbf{g} \times \operatorname{grad} \varpi] > 0. \tag{5}$$

If this condition is violated, one can always choose a virtual displacement in such a way that it will decrease the total energy. The configuration is thus unstable and this is known as the *baroclinic instability*. The vector condition (5), together with (4), shows that this instability arises whenever the specific angular momentum $\varpi^2 \Omega$ decreases towards the equator on a surface of constant specific entropy. This criterion is more general than the Rayleigh criterion, which, of course, it includes. It shows also that a stable density stratification permits certain rotation laws which depend on the z coordinate, thereby violating the Taylor-Proudman theorem.

1.2. SHEAR INSTABILITY

Unfortunately, it has not yet been possible to extend the stability criterion given above to more general, non-axisymmetric, virtual displacements. However, the instability which is likely to be most prominent in a fluid which is not in solid body rotation when perturbed by non-axisymmetric disturbances, is the so-called shear instability.

In a plane parallel shear flow, a sufficient condition for stability is that the velocity profile present no inflection point. This theorem has also been established by Rayleigh, and its counterpart for a cylindrical flow is that the expression

$$\frac{d}{d\varpi}\left[\frac{1}{\varpi}\frac{d}{d\varpi}(\varpi^2 \Omega)\right]$$

does not change sign in the domain considered. What happens when this expression vanishes is not clear, although it is likely that dynamical instability will appear for a large variety of velocity profiles. But this is of rather academic interest since laboratory experiments and some theoretical investigations have shown that the viscosity is able to destabilize any flow, even if it is claimed to be stable according to Rayleigh's theorem above, provided that the Reynolds number VL/ν exceeds some critical value which is of order 10^3. (V and L are respectively typical values for the velocity and the dimension of the domain and ν is the kinematic viscosity).

Here again the flow may be stabilized by a (stable) density stratification, provided that the Richardson number R_i exceeds some critical value:

$$R_i = \frac{g}{C_P}\frac{dS}{dz}\bigg/\left(\frac{dv}{dz}\right)^2 > \tfrac{1}{4}. \tag{6}$$

A similar condition probably applies also to differential rotation, the velocity gradient being replaced by $\varpi \operatorname{grad} \Omega$, and only the components of $\mathbf{g}$ and $\operatorname{grad} S$ parallel to the gradient of Ω being retained in the criterion:

$$\cos^2 \alpha \, \frac{g}{C_P}\frac{dS}{dr} > \tfrac{1}{4}(\varpi \operatorname{grad} \Omega)^2 \tag{7}$$

(here α is the angle between $\mathbf{g}$ and $\operatorname{grad} \Omega$).

2. Destabilization Through Thermal Diffusion

As we have just seen, a suitable density stratification can stabilize flows which would otherwise be unstable. But this situation changes drastically, at least in a homogeneous fluid, if the perturbations are not constrained to be adiabatic. The reason for this is that thermal diffusion smoothes out the temperature fluctuations, and hence the density perturbations since on the time scale considered here the medium is in almost perfect pressure equilibrium. As a result, the stabilizing buoyancy forces are weakened and, in some cases, can no longer prevent the muted instability.

2.1. Axisymmetric Instability

In a rotating star, the effect of smoothing the density perturbation is to decrease the quadratic term Q_1 in $\delta^2\varepsilon$ (Equation (1)) to a point where it becomes negligible compared to Q_2. Instability then sets in whenever Q_2 fails to be a definite positive quadratic form, and this happens, as we have already seen, when

$$\frac{\partial}{\partial \varpi}(\varpi^2\Omega)^2 < 0 \quad \text{or} \quad \frac{\partial}{\partial z}(\varpi^2\Omega)^2 \neq 0. \tag{3a, b}$$

Goldreich and Schubert (1967) and Fricke (1968) were the first to describe this effect in the astrophysical context, but credit should also be given to Yih (1961) for having called attention to the destabilizing role of thermal conduction in a cylindrically rotating flow. The question of the growth rate of this instability has not yet been settled, although in the linear phase of its development it is tied to the thermal time scale. James and Kahn (1970) endeavoured to follow the perturbation into the nonlinear regime, where its behaviour must bear much resemblance with that of thermohaline convection.

In thermohaline convection, an unstable salt gradient is stabilized by a stable temperature – (and thus density) – gradient. As Goldreich and Schubert pointed out, the situation is very similar to that of a rotating fluid of low Prandtl number* since, in both cases, the thermal diffusivity is much larger that either the molecular diffusivity of salt or the viscous diffusivity of angular momentum.

For a comprehensive account of thermohaline convection, we refer to Spiegel (1972). Laboratory experiments and numerical investigations show that the dominant unstable modes are of high horizontal wave number; they evolve in what is called the salt fingers. As those fingers reach an appreciable vertical size, they seem to become collectively unstable and turn into layers with discontinuities in concentration, such as those observed in the oceans. Finally, it is found that such layering can be destroyed by sufficiently strong turbulence.

It is very likely that the Goldreich-Schubert-Fricke instability will also result in a similar layering of angular momentum, which in turn, will then become eligible for shear instability, if it has not been so before.

---

* The Prandtl number is the ratio of the kinematic viscosity to the thermal diffusivity $\sigma = \nu/\kappa$.

2.2. Shear instability

Such a breakdown of the density stabilization is also expected in the shear flow. In the Earth's atmosphere, turbulent motions are observed in regions for which the Richardson criterion predicts strong stability. In order to explain this, Townsend (1958) took thermal diffusion into account in the energy balance of a turbulent shear flow. He showed that this effect lowers considerably the critical Richardson number if the radiative cooling time of the perturbations is small enough, so that

$$t_{\text{cool}} \left|\frac{dv}{dz}\right| < \sim \tfrac{1}{6}.$$

Townsend introduces then a new non-dimensional number, the flux Richardson number, which must be larger than unity for the turbulence to be prevented by the density stratification:

$$\frac{g}{C_P}\frac{dS}{dz} t_{\text{cool}} \bigg/ \left|\frac{dv}{dz}\right| > \sim 1. \tag{8}$$

Moore and Spiegel (1964) have argued that one should take for t_{cool}, the cooling time corresponding to the smallest scale which is optically thick in the medium. This length is of the order of a centimeter in stellar interiors and the application of Townsend's criterion would then lead to critical rotation laws with exceedingly small gradients.

In Townsend's treatment, the cooling time is that associated with the integer scale of the turbulence, which is probably much larger than the one centimeter above. Unfortunately, in the absence of direct observations one has to resort to some guesswork about it. It seems however reasonable to take for it the shortest length l whose Reynolds number still remains supercritical:

$$\frac{l^2}{v}\left|\frac{dv}{dz}\right| = R_{\text{crit}}, \tag{9}$$

where R_{crit} is the critical Reynolds number ($\sim 10^3$) for the given velocity profile. Among all the scales present in the spectrum of the turbulence, this one is likely to stand out: it has the largest growth-rate on the source side of the spectrum, where energy is fed into the turbulence from the unstable laminar flow.

Using the radiative cooling time corresponding to this scale, one finds that the critical Richardson number is increased to a value of the order $(\sigma R_{\text{crit}})^{-1}$ (see footnote on p. 189), typically 10^3 or 10^4 in stellar interiors. This leads to the following estimation of the gradient of angular velocity which cannot be stabilized by a given density stratification:

$$(\varpi \, \text{grad}\, \Omega)^2 > (\sigma R_{\text{crit}}) \cos^2 \alpha \, \frac{g}{C_p}\frac{dS}{dr}$$

or
$$(\varpi \text{ grad} \log \Omega)^2 > (\sigma R_{\text{crit}}) \cos^2 \alpha \left(\frac{g}{\Omega^2 H_p}\right) (\nabla_{\text{AD}} - \nabla_{\text{RAD}}) \tag{10}$$

with the usual notations.

A perturbation of the scale assumed above has a growth time of order $t = l^2/v = R_{\text{crit}} |\varpi \text{ grad} \Omega|^{-1}$, only three or four orders of magnitude larger than the dynamical time scale. We may thus conclude that gradients of angular velocity which are larger than those indicated by condition (10) will be wiped out in time which is short compared to any time characterizing the thermal or nuclear evolution of the star.

3. Stabilization Through Chemical Composition Gradients

We have just seen how efficiently thermal diffusion suppresses the stabilizing effect of a density stratification. But this is true only as long as the fluid is homogeneous. If the density gradient is sustained by a stable gradient of chemical composition, the density fluctuations cannot be cancelled completely by thermal diffusion and the remaining buoyancy forces may still be strong enough to prevent the instability.

3.1. Axisymmetric Instability

In the case of axisymmetric disturbances, this can be seen by deriving the corresponding second variation $\delta^2\varepsilon$ of the total energy, as in the case of adiabatic perturbations. The star is assumed to be in overall secular stability, so that the first variation $\delta\varepsilon$ is again zero. One has to assume also that the perturbation evolves slowly enough so that the Eulerian variations of pressure and temperature can be neglected. The variation of density is then due only to the advection of matter of varying molecular weights

$$\delta\varrho/\varrho = -\delta\mathbf{r} \cdot \text{grad} \log \mu$$

and the quadratic term Q_1 becomes:

$$Q_1 = \delta\mathbf{r} \cdot (-\text{grad} \log \mu)(-\mathbf{g}) \cdot \delta r.$$

It is not possible to derive a general secular stability criterion, as was done previously for the dynamical stability, since the four vectors involved in the tensor $\mathfrak{M}$ are no longer related by an equation similar to the baroclinic one (4). (For a perfect gas, $\text{grad} \log \mu/T$ is still related to the three other vectors, but $\text{grad} \log \mu$ is not). However, one can proceed along the following lines.

Since the stabilization comes from the quadratic term Q_1, let us take the most favorable case, for which the vectors $\mathbf{g}$ and $\text{grad} \log \mu$ are colinear: Q_1 is then positive definite, i.e. positive except for $\delta\mathbf{r}$ perpendicular to those vectors where it vanishes. In this case, it is easy to show that the total quadratic form $\delta\mathbf{r} \cdot \mathfrak{M} \cdot \delta\mathbf{r}$ is definite positive if and only if:

$$\left(\frac{1}{\varpi^3} \text{grad}(\varpi^2 \Omega)^2 \times \text{grad} \varpi\right)^2$$

$$< 4 \left(-|g| \frac{d}{d\zeta} \log\mu \right) \left(\mathbf{1}_\zeta \times \frac{1}{\varpi^3} \operatorname{grad}(\varpi^2 \Omega)^2 \right) \cdot (\mathbf{1}_\zeta \times \operatorname{grad} \varpi), \quad (12)$$

where $\mathbf{1}_\zeta$ is the unit vector in the vertical direction, i.e. $\mathbf{1}_\zeta = -\mathbf{g}/|g|$. If the angular velocity is a function of the vertical coordinate ζ only, or if the horizontal variation of Ω can be neglected compared to the vertical variation, this awkward condition takes a much simpler form:

$$-|g| \frac{d}{d\zeta} \log\mu > \tfrac{1}{4} \left(\varpi \frac{d\Omega}{d\zeta} \right)^2. \quad (13)$$

One may therefore conclude, as did Goldreich and Schubert, that a sufficiently strong gradient of molecular weight is capable of suppressing the instability named after these authors.

3.2. Shear Instability

Such a μ-gradient will also inhibit the shear instability; if, in the derivation of the Richardson criterion, one assumes the perturbations to be isothermal instead of adiabatic, the gradient of molecular weight takes the place of the density gradient. Transposed to a rotating fluid, the Richardson criterion for stability then becomes:

$$-\cos^2\alpha |g| \frac{d}{d\zeta} \log\mu > \tfrac{1}{4} (\varpi \operatorname{grad}\Omega)^2 \quad (14)$$

which is identical to condition (13) above, but with no restrictive assumption about the rotation law (α is again the angle between grad Ω and $\mathbf{g}$; the μ-gradient is supposed to be purely vertical).

4. The Evolution of Rotating Stars

Few attempts have been made so far to follow the detailed history of a rotating star (see e.g., Kippenhahn et al., 1970), although it has been known at least since Eddington's work that rotation can play an important role in the evolution of a star. True, it is only very recently that the destabilizing role of thermal diffusion has been recognized and this explains why most evolutionary calculations take the rotational instabilities into account only very crudely, and why some neglect them completely (e.g. Sakurai, 1972).

In the absence of more detailed calculations, one can hardly do better than sketch roughly what, with some likelihood, appears to be the evolution of a rotating star.

If the star, when it first reaches the main sequence, is homogeneous as the result of convective mixing during the Hayashi phase, it will settle into a cylindrical rotation state due to the interplay between the two instabilities described above (Goldreich-Schubert-Fricke's and shear). This adjustment takes only a very short time, of the order of 10^3 or 10^4 rotation periods.

In the case of a late-type star, the rotation law is then entirely determined by the boundary conditions at the bottom of the outer convection zone*. If, as it is now believed after Schatzman (1962), the rotation of the convection zone is slowed down by a stellar wind sustained by strong magnetic fields, the rotation of the radiative interior will adjust to it in the short time already mentioned. It is only after this braking mechanism has become very weak, probably when its (negative) e-folding time has become comparable with the nuclear evolution time, that the motions in the radiative core of the star have damped sufficiently so that a gradient of molecular weight can be established. After this time, the star evolves inhomogeneously and its core is shielded from further rotational instability. After the main-sequence phase, the contracting core will conserve its angular momentum and will spin much faster than the expanding envelope.

In this scheme, the present Sun would have a core rotating with the angular velocity determined by the angular momentum it had at the beginning of its inhomogeneous phase. Since the Sun has probably not slowed down subsequently by more than a factor of two or three, it is not likely that it possesses the fast spinning core needed to support Dicke's theory (1964), or to lower the calculated neutrino flux to the presently observed level (see Demarque's contribution at this symposium).

In an early type star, the rotation law $\Omega(\varpi)$ is similarly determined by the convective core, but only inside the cylinder tangent to it at its equator. Outside this region, the situation is more complex; it is probably the Eddington-Sweet circulation which plays the main role in redistributing angular momentum, within the limit fixed by condition (10) which does not tolerate departures from solid rotation much larger than a few per cent. The gradients of molecular weight built by this circulation (Mestel's μ-barriers, 1953) and those later left behind by the receding convective core are probably strong enough to shield the central regions from the rotational instabilities as in the case of a low mass star.

All this has clearly to be checked by detailed evolutionary calculations. The only excuse for the present speculations is that they may stimulate such much needed investigations.

Acknowledgements

I wish to thank Prof. R. Van der Borght and his colleagues of Monash University, where this review has been prepared, for their generous hospitality. I am also indebted to Prof. R. H. Koch for having critically read the manuscript and greatly improved the English.

References

Dicke, R. H.: 1964, *Nature* **202**, 432.
Eliassen, A. and Kleinschmidt, E.: 1957, *Handbuch der Physik* **48**, 64.

---

* Should the differential rotation imposed by the convection zone be strong enough to trigger the shear instability according to condition (10), the angular velocity of the radiative interior would probably remain in neutral equilibrium and connect with the angular velocity in the convection zone through a turbulent Ekman layer.

Fricke, K. J.: 1968, *Z. Astrophys.* **68**, 317.
Fricke, K. J. and Kippenhahn, R.: 1972, *Ann. Rev. Astron. Astrophys.* **10**, 45.
Goldreich, P. and Schubert, G.: 1967, *Astrophys. J.* **150**, 571.
Greenspan, H. P.: 1968, in G. K. Batchelor and J. W. Miles (ed.), *The Theory of Rotating Fluids*, Cambridge University Press, London, p. 2.
Howard, L. N., Moore, D. W., and Spiegel, E. A.: 1967, *Nature* **214**, 1297.
James, R. A. and Kahn, F. D.: 1970, *Astron. Astrophys.* **5**, 232.
Kippenhahn, R., Meyer-Hofmeister, E., and Thomas, H. C.: 1970, *Astron. Astrophys.* **5**, 155.
Lebovitz, N. R.: 1967, *Ann. Rev. Astron. Astrophys.* **5**, 465.
Ledoux, P.: 1958, *Handbuch der Physik* **51**, 605.
Mestel, L.: 1953, *Monthly Notices Roy. Astron. Soc.* **113**, 716.
Mestel, L.: 1965, *Stars and Stellar Systems* **8**, 465.
Mestel, L.: 1970, Circular Letter No. 10, IAU Comm. 35.
Moore, D. W. and Spiegel, E. A.: 1964, *Astrophys. J.* **139**, 48.
Ostriker, J. P.: 1970, in A. Slettebak (ed.), 'Stellar Rotation', *IAU Colloq.* **4**, 147.
Sakurai, T.: 1972, *Publ. Astron. Soc. Japan* **24**, 153.
Schatzman, E.: 1962, *Ann. Astrophys.* **25**, 18.
Spiegel, E. A.: 1972, *Ann. Rev. Astron. Astrophys.* **10**, 261.
Spiegel, E. A. and Zahn, J. P.: 1970, *Comments Astrophys. Space Phys.* **2**, 178.
Strittmatter, P. A.: 1969, *Ann. Rev. Astron. Astrophys.* **7**, 665.
Townsend, A. A.: 1958, *J. Fluid Mech.* **4**, 361.
Yih, C.-S.: 1961, *Phys. Fluids* **4**, 806.

DISCUSSION

(*Note:* Several questions have been raised in the discussion by Drs Demarque, Schatzman and Tayler concerning the rotational evolution of the Sun. Dr Zahn has preferred to include his answers in the paper itself, whose last paragraph has been rewritten for this purpose.)

Tayler: Could I make a comment on a problem that Dr Zahn has not actually spoken about. That is the interference of rotation with convection, which of course is buried in his criteria somewhere. If one wants a simple sentence to bear on this, rotation will interfere with convection, if the rotation time is short compared to the lifetime of the convective element.

Zahn: According to the criterion (2) you have always some displacement which should be unstable, even if you have a stabilizing rotation.

Tayler: There will always be some unstable perturbation. I said it will interfere with convection, I did not say it would suppress it.

Zahn: Yes.

Rodgers: It seems to me that the only objects I can think of straight away which are regarded as purely radial pulsators and which may have relatively high observed rotational velocities are the stars which get into the instability zone through the McCray mechanism, as blue stragglers. We are talking, if we want to be specific, about one star called HD 6870, where we have $V \sin i$ typically around 120 km per second. Prof. Eggen has observed it as a low amplitude variable of the order of a few hundredths of a magnitude but with typical δ Scuti characteristics. What I want to know is what do you say about the effects of rotation on radial pulsation?

Zahn: This was a part of Prof. Ledoux's topic. As he has mentioned the radial oscillation will also be affected by rotation, there will be a shift in frequency, a slight one, but that is all.

Ledoux: It is difficult to define a purely radial oscillation in something that is rotating pretty fast, but you still have some kind of pseudo radial mode which is usually not too different in period from the radial mode.

Zahn: But the blue stragglers raise another question. How do they occur at all? You remember that Prof. Mestel found that the μ-gradients will prevent mixing in the star. It may well be that if there is a stronger rotation present those μ-gradients will be broken or something like that happens. Has someone looked in detail at that?

Cox: Let me ask the same question as Dr Rodgers. Is it true that rotation will decrease the amplitude of purely radial pulsation?

Zahn: To tell something about the amplitude you have to do nonlinear calculations. The amplitude may be reduced because the Coriolis force will try to make the oscillation more horizontal. But you would have to look in detail, I have no real answer.

Vardya: This is similar to what Dr Demarque asked earlier. A μ-gradient will be built up also as you go to the core because of ionisation. How much of a μ-gradient do you want? This will occur when the star is homogeneous.

Zahn: The μ-gradient which enters in the instability criteria (13) and (14) is that caused by a genuine variation of chemical composition. The state of ionisation adjusts too fast to a change in pressure or temperature to play here any role. Moreover, the resulting μ-gradient would be destabilizing (μ-increasing outwards).

PERTURBATION OF THE NON-RADIAL OSCILLATIONS OF A GASEOUS STAR BY AN AXIAL ROTATION, A TIDAL ACTION OR A MAGNETIC FIELD

P. SMEYERS

Astronomisch Instituut, Universiteit Leuven, Belgium

Abstract. The study of the linear and adiabatic oscillations of a gaseous star gives rise to an eigenvalue problem for the pulsation σ, if perturbations proportional to $e^{i\sigma t}$ are considered. In the presence of a rotation, a tidal action or a magnetic field, the equations are not separable in spherical coordinates. To get approximate expressions for the influence of these factors on the non-radial oscillations of a star, the author and his collaborators J. Denis and M. Goossens have used a perturbation method (Smeyers and Denis, 1971; Denis, 1972; Goossens, 1972; Denis, 1973). Their procedure corresponds to a generalization of the method proposed by Simon (1969) to study the second order rotational perturbation of the radial oscillations of a star.

Two types of perturbations are taken into account: volume perturbations due to the local variations of the equilibrium quantities and to the presence of a supplementary force in the equation of motion (Coriolis force, Lorentz force); surface perturbations related to the distortion of the equilibrium configuration and to the change of the condition at the surface in the presence of a magnetic field. The resulting expressions are accurate up to the second order in the angular velocity in the case of a rotational perturbation, to the third order in the ratio of the mean radius of the primary to the distance of the secondary in the case of a tidal perturbation, and to the second order in the magnetic field in the case of a perturbing magnetic field. These expressions can in principle be applied to any mode.

Numerical results have been obtained for a homogeneous model and for a polytropic model $n=3$. In particular, the splitting of the frequencies of the fundamental radial mode and of the f-mode belonging to $l=2$ and $m=0$ has been studied for the critical value of γ, in the case of a component of a synchronously rotating binary system.

References

Denis, J.: 1972, *Astron. Astrophys.* **20**, 151.
Denis, J.: 1973, Doctoral Thesis, Louvain, Belgium.
Goossens, M.: 1972, *Astrophys. Space Sci.* **16**, 386.
Simon, R.: 1969, *Astron. Astrophys.* **2**, 390.
Smeyers, P. and Denis, J.: 1971, *Astron. Astrophys.* **14**, 311.

DISCUSSION

Zahn: When you consider the rotational and tidal bulge, do you consider only the axis of the tidal component, because the tidal bulge component is perpendicular to the axis of rotation and this has the same symmetry as the rotational case. So do you consider the axisymmetrical components only?

Smeyers: We take into account both the distortions caused by rotation and by tidal action. In the development of the tidal potential, we keep the Legendre polynomials $P_2(\cos\Theta)$, $P_3(\cos\Theta)$ and $P_4(\cos\Theta)$, Θ being the angle between the direction from the origin to the companion and the direction to the point considered.

MASS-RATIO DETERMINATION IN CONTACT BINARIES

J. B. HUTCHINGS

Dominion Astrophysical Observatory, Victoria, B.C., Canada

Abstract. In the wake of recent theoretical work on contact systems (e.g. Whelan, 1972; Biermann and Thomas, 1972; Lucy, 1968), it is of importance to determine fundamental data from observations. This has been done recently by several groups in analysing light curves (Mochnacki and Doughty, 1972; Hutchings and Hill, 1973; Wilson and Devinney, 1973), and it is found that shapes, temperature differences and distributions, and mass-ratios, can be determined in many cases. However, where spectroscopic data are also available, the mass-ratios are not always in agreement. Using the photometric models, it is possible (Hutchings, 1973) to calculate the distortion of line profiles resulting (primarily) from the non-uniform brightness over the component stars in these systems. This distortion leads to the characteristically observed 'square' velocity curves for the systems (e.g. Binnendijk, 1967). Correction for the effect in most cases (a) resolves the mass-ratio discrepancy and (b) leads to better estimates for the masses. The faintness of most contact systems makes detailed spectroscopy difficult, but there appears to be a need for further work in the directions outlined here to improve the fundamental data available on them. These results should also be borne in mind in inspecting previous work on contact binaries.

References

Biermann, P. and Thomas, H. C.: 1972, *Astron. Astrophys.* **16**, 60.
Binnendijk, L.: 1967, *Publ. Dominion Astrophys. Obs.* **13**, 27.
Hutchings, J. B.: 1973, *Astrophys. J.* **180**, 501.
Hutchings, J. B. and Hill, G.: 1973, *Astrophys. J.* **179**, 539.
Lucy, L. B.: 1968, *Astrophys. J.* **153**, 877.
Mochnacki, S. and Doughty, N. A.: 1972, *Monthly Notices Roy. Astron. Soc.* **156**, 243.
Whelan, J. A. J.: 1972, *Monthly Notices Roy. Astron. Soc.* **156**, 115.
Wilson, R. E. and Devinney, E. J.: 1973, *Astrophys. J.* **182**, 539.

INDEX OF NAMES

Bold printed numbers refer to papers, normally printed numbers to discussion remarks

Aizenman, M. L., 73, 78

Bahner, K., v
Bell, R. A., **57**, 57, 60, 87, 99
Bessell, M. S., 36, 60, **63**, 64, 65, 66, 78, 79, 86, 108
Breger, M., 34, 59, **61**, 61, 64, 65, 66, 71, 83
Buscombe, W., 35, 47, 53, **87**, 87, 98, 173, 179

Cogan, B. C., 60
Cox, A. N., **39**, 46, 47, 48, 52, 53, 56, **59**, 59, 60, 67, 68, **73**, 78, 102, 105, 108, 195

Demarque, P., 35, 105, 106, 111, 113
Duthie, J. G., **129**

Eggen, O. J., **35**, 35, 36, 48

Faulkner, D., 113
Feast, M. W., 36, **93**, 98, 99, 100

Graham, J. A., 70, **107**, 108

Hutchings, J. B., 87, **127**, **199**
Hyland, H. R., 86, 100

Iben, I., Jr., **3**, 33, 34, 36, 46, 47, 56, 60, 66, 73, 78, 101, 102, 105, 108, 111, 113, 114
Irwin, J. B., 47, 61, 71, 86

Katz, J., 113
Kemp, J. C., 67, 102, **179**, 179, 180
King, D. S., **59**
Kochhar, R. K., **181**

Langford, W. R., **65**
Ledoux, P., **135**, 173, 194
Lesh, J. R., 71

McNamara, D. H., 56, 64, **65**, 65, 66
Maeder, A., **81**, 83, **109**, 111, 173

Malone, R., **113**
Mavridis, L. N., v, 98

Osmer, P. S., **85**, 86

Papaloizou, J. C. B., **77**
Parsons, S. B., **55**, 56, 86
Przybylski, A., 86, 100

Renaud, B., **129**
Rodgers, A. W., 33, 34, 35, 36, 46, 47, 48, 53, 56, 57, 60, 61, 66, 67, 68, 71, 83, 99, 100, 108, 194
Rufener, F., **81**

Salpeter, E. E., **113**, 113, 114
Sargent, W. L., 78
Savedoff, M. P., 68, **129**
Schatten, K. H., **175**, 177
Schatzman, E., 56, 59, 67, 68, 102, **117**, 123
Shobbrook, R. R., **67**, 68, **69**, 70, 71
Sinvhal, S. D., 65, 68
Smeyers, P., **197**, 198
Stobie, R. S., 34, 47, **49**, 53, 60, 65, **67**, 67, 68, 78

Tabor, J. E., **59**, **73**
Tayler, R. J., 34, 57, 60, **77**, 78, 79, 105, 173, **177**, 177, 180, 194
Trehan, S. K., **181**

Underhill, A. B., 83, 87

Van Woerden, H., 78
Vardya, M. S., 99, 100, 195

Warner, B., 68, **125**, 173
Webster, B. L., 114, **123**
Wood, P. R., **101**, 101, 102

Zahn, J.-P., 47, 48, 60, 105, **185**, 194, 195, 197